Musical Instruments

Musical Instruments

Acoustics and Vibration

Editor

Lamberto Tronchin

MDPI • Basel • Beijing • Wuhan • Barcelona • Belgrade • Manchester • Tokyo • Cluj • Tianjin

Editor
Lamberto Tronchin
Department of Architecture,
University of Bologna
Italy

Editorial Office
MDPI
St. Alban-Anlage 66
4052 Basel, Switzerland

This is a reprint of articles from the Special Issue published online in the open access journal *Applied Sciences* (ISSN 2076-3417) (available at: https://www.mdpi.com/journal/applsci/special_issues/Musical_Instruments).

For citation purposes, cite each article independently as indicated on the article page online and as indicated below:

LastName, A.A.; LastName, B.B.; LastName, C.C. Article Title. *Journal Name* **Year**, *Article Number*, Page Range.

ISBN 978-3-03936-613-2 (Hbk)
ISBN 978-3-03936-614-9 (PDF)

Contents

About the Editor

Lamberto Tronchin has served as Associate Professor at University of Bologna since 2011, where he is active in applied acoustics and energy efficiency research. His interests mainly regard room acoustics, where he has worked on developing new methods to measure acoustic quality in rooms, the design of theatres and auditoria, and the characterization of musical acoustics, where he has developed new vibro-acoustic parameters (IAR) and emulated the nonlinear sound behavior of musical instruments by means of Volterra series. With respect to energy efficiency, his research involves the study of new materials for improved energy efficiency of buildings. He is involved in both EU (POR FESR 2014–2020) and national (PRIN2015) projects. He is author of more than 200 papers and has served as plenary lecturer at numerous international congresses and institutions. He is also inventor of an international patent belonging to University of Bologna, namely "Method for artificially reproducing an output signal of a non-linear time invariant system". He is President of AES—Italian Section.

Editorial

Special Issue on Musical Instruments: Acoustics and Vibration

Lamberto Tronchin

Department of Architecture, University of Bologna, Via dell'Università 50, 47521 Cesena, Italy; lamberto.tronchin@unibo.it

Received: 6 May 2020; Accepted: 6 May 2020; Published: 9 May 2020

1. Introduction

The sound characteristics of musical instruments have been constantly growing in importance. Consequently, several congresses, workshops, and conferences have been organized in the last ten years. The studies on musical instruments, their mechanical behavior, sound emission, and characteristics started thousands of years ago, and among the physicists and mathematicians that addressed this matter, we should at least remember Leonardo da Vinci, with his experimental water organ, and Ernst Chladni, who discovered the nodal patterns on rigid surfaces, such as soundboards. The growing awareness of our intangible cultural heritage and the need to better understand our roots in the field of music have contributed to increasing the efforts to extend our knowledge in this field, defining new physical parameters, extending the analysis to other musical instruments, and developing new methods to synthesize sound from musical instruments using a simple keyboard.

These motivations led us to the proposal of a special issue called "Musical Instruments: Acoustics and Vibration" since we believe in the importance of musical acoustics within modern acoustics studies. In total, 13 papers were submitted and 8 of them were published, with an acceptance rate of 61.5%. Among all the papers published, one of them was classified as a review paper, while the rest were classified as research papers. According to the number of papers submitted, and the specificity of the musical acoustics branch within acoustics, it can be affirmed that this is a trendy topic in the scientific and academic community and this special issue on "Musical Instruments: Acoustics and Vibration" aims to be a future reference for the research that is to be developed in the next few years.

2. Musical Instruments: Acoustics and Vibration

Human beings started to play early musical instruments in the Neanderthal age [1], a fact that helps us to understand the importance of music for the world.

The sound characteristics of musical instruments, as well as their vibrational behavior, represent one of the most important and fascinating fields of acoustics, or even of applied physics.

This aspect is sometimes neglected (or at least not investigated enough) during the restoration of ancient masterpieces, even though it is well known that their sound production is something without equal and of inestimable value.

Following this concept, this special issue aimed to contribute to the knowledge of the acoustics of musical instruments. This goal was reached by proposing (or applying) new methods for characterizing the acoustics of musical instruments, by presenting studies on some specific art pieces, or by trying to illustrate some applications in sound synthesis.

The paper by Turk et al. [1] gives an interesting review of the historical debate about the findings of the "Neanderthal musical instrument" from the "Divje Babe I Cave" (Slovenia), one of the most ancient finds related with musical instruments, at least in Europe. This paper gives a proper idea about the ancient origin of this matter.

The two papers by Tronchin et al. [2,3] analyze completely different musical instruments. Starting from the definition of a new vibro-acoustical parameter called the intensity of acoustic radiation (IAR) [2], which was initially proposed for kettledrums, the studies were carried out to contribute to the knowledge of special and rare musical instruments. The first paper reports the results of both the modal analysis and IAR measured in a thar, a sithar, and a santoor, three important Persian musical instruments [3]. These outcomes give an idea of their behavior in response to increasing demand for knowledge of those musical instruments. The second paper describes the outcome of an experimental analysis carried out on a carabattola, a largely unknown ethnic Italian musical instrument, which used to be played in the Romagna region until the Second World War [4]. The analysis includes modal analysis and IAR measurements. It gives a unique contribution to the knowledge of this unique instrument.

The paper by Ibáñez-Arnal et al. [5] shifts the discussion to the physical properties of the material utilized for the realization of musical instruments, focusing on the carbon fiber reinforced epoxy (CFRE) prepregs, which could be used for new prototypes of new musical instruments. Undoubtedly, the physical characteristics of the materials strongly contribute to the overall assessment of the sound quality of the instruments.

The other papers focus on the application of the physics of musical instruments in the emulation of their sound production, especially during synthesis or recording. The paper by Moore [6] proposes a method for analyzing the dynamic range of sounds and music, whilst the paper by Papetti et al. [7] applies the outcomes of their previous studies into a new audio-tactile piano sample library, which is useful for real-time performances.

The last two papers analyze some specific aspects of this intriguing matter, especially from the signal processing perspective. In their paper, Jiang et al. [8] analyze the timbre perception features in musical motifs, whilst Ziemer and Plath [9] describe a method for simulating sound radiation using a microphone and loudspeaker array, going into detail about the necessary signal processing; the techniques used in both of these papers could be implemented when analyzing the non-linear components of the sound quality of musical instruments [10,11].

3. Conclusions

All the results presented and published in this special issue suggest that the acoustics and vibration of musical instruments is a relevant and popular topic in the scientific community. The results reported by all the authors increase the knowledge in this subject and contribute to a further understanding of this matter. This issue could become a starting point for further developments in the area of the physics of musical instruments.

Funding: This research was funded by Regione Emilia Romagna POR-FESR 2014-20 "SIPARIO" grant number PG/2018/632038.

Acknowledgments: The success of this special issue is strongly related to the huge work and the great contributions of all the authors. Furthermore, we acknowledge the hard work and the professional support of the reviewers and the editorial team of Applied Sciences. We are extremely grateful to all the reviewers involved in the issue for their time and their knowledge. We thank the assistant editors from MDPI that collaborated with us for their tireless support. We hope that the editorial process, starting from the submission and focusing on the review, was appreciated by all the authors, despite the final decisions. The real value of the time and the work spent in this process is found in the help provided to the authors to improve their papers.

Conflicts of Interest: The author declares no conflict of interest.

References

1. Turk, M.; Turk, I.; Otte, M. The Neanderthal Musical Instrument from Divje Babe I Cave (Slovenia): A Critical Review of the Discussion. *Appl. Sci.* **2020**, *10*, 1226. [CrossRef]
2. Tronchin, L. Modal analysis and intensity of acoustic radiation of the kettledrums. *J. Acoust. Soc. Am.* **2005**, *117*, 926–933. [CrossRef] [PubMed]

3. Tronchin, L.; Manfren, M.; Vodola, V. Sound Characterization through Intensity of Acoustic Radiation Measurement: A Study of Persian Musical Instruments. *Appl. Sci.* **2020**, *10*, 633. [CrossRef]
4. Tronchin, L.; Manfren, M.; Vodola, V. The Carabattola—Vibroacoustical Analysis and Intensity of Acoustic Radiation (IAR). *Appl. Sci.* **2020**, *10*, 641. [CrossRef]
5. Ibáñez-Arnal, M.; Doménech-Ballester, L.; Sánchez-López, F. A Study of the Dynamic Response of Carbon Fiber Reinforced Epoxy (CFRE) Prepregs for Musical Instrument Manufacturing. *Appl. Sci.* **2019**, *9*, 4615.
6. Moore, A. Dynamic Range Compression and the Semantic Descriptor Aggressive. *Appl. Sci.* **2020**, *10*, 2350. [CrossRef]
7. Papetti, S.; Avanzini, F.; Fontana, F. Design and Application of the BiVib Audio-Tactile Piano Sample Library. *Appl. Sci.* **2019**, *9*, 914. [CrossRef]
8. Jiang, W.; Liu, J.; Zhang, X.; Wang, S.; Jiang, Y. Analysis and Modeling of Timbre Perception Features in Musical Sounds. *Appl. Sci.* **2020**, *10*, 789. [CrossRef]
9. Ziemer, T.; Plath, N. Microphone and Loudspeaker Array Signal Processing Steps towards a "Radiation Keyboard" for Authentic Samplers. *Appl. Sci.* **2020**, *10*, 2333. [CrossRef]
10. Tronchin, L. The Emulation of Nonlinear Time-Invariant Audio Systems with Memory by Means of Volterra Series. *J. Audio Eng. Soc.* **2012**, *60*, 984–996.
11. Tronchin, L.; Coli, V.L. Further investigations in the emulation of nonlinear systems with Volterra series. *J. Audio Eng. Soc.* **2015**, *63*, 671–683. [CrossRef]

applied sciences

Review

The Neanderthal Musical Instrument from Divje Babe I Cave (Slovenia): A Critical Review of the Discussion

Matija Turk [1,2,*], **Ivan Turk** [3] **and Marcel Otte** [4]

1   ZRC SAZU Institute of Archaeology, Novi trg 2, 1000 Ljubljana, Slovenia
2   National Museum of Slovenia, Prešernova 20, 1000 Ljubljana, Slovenia
3   Lunačkova 4, 1000 Ljubljana, Slovenia; ivan.turk.46@gmail.com
4   Université de Liège, 7, Place du XX Août, Bât. A1, 4000 Liège, Belgium; marcel.otte@ulg.ac.be
*   Correspondence: matija.turk@zrc-sazu.si

Received: 28 November 2019; Accepted: 26 January 2020; Published: 12 February 2020

Abstract: The paper is a critical review of different evidence for the interpretation of an extremely important archaeological find, which is marked by some doubt. The unique find, a multiple perforated cave bear femur diaphysis, from the Divje babe I cave (Slovenia), divided the opinions of experts, between those who advocate the explanation that the find is a musical instrument made by a Neanderthal, and those who deny it. Ever since the discovery, a debate has been running on the basis of this division, which could only be closed by similar new finds with comparable context, and defined relative and absolute chronology.

Keywords: Palaeolithic; Mousterian; Neanderthals; musical instrument; Divje babe I

1. Introduction

Discoveries that shed light, directly or indirectly, on the spiritual life of Neanderthals always attract great attention from the professional and lay public. One such find was unearthed in 1995 in Mousterian level D-1 (layer 8a), as a result of long-lasting (1979–1999) excavations in the Palaeolithic cave site of Divje babe I (DB) in western Slovenia, conducted by the ZRC SAZU Institute of Archaeology from Ljubljana. It was a left femur diaphysis, belonging to a one to two-year-old cave bear cub with holes (inventory no. 652), which resembled a bone flute (Figure 1). The object was found cemented into the breccia in the immediate vicinity of Neanderthal hearth, placed into a pit [1,2].

The excavation leader, I. Turk, proposed two possible explanations soon after its discovery: An artefact or a pseudo-artefact in the form of a gnawed and teeth-pierced femur diaphysis [1]. According to the first explanation, this find would be the oldest musical instrument [2–10]. The main surprise was not the great age of the find (at first 45,000 years, later 50,000–60,000 years), determined with ^{14}C AMS, U/Th, and ESR on accompanying finds of charcoal, cave bear bones and teeth [8,9,11], but its undeniable attribution to Mousterian culture, i.e., Neanderthals. As such, it would represent significant evidence for existence of musical behaviour, long before the spread of anatomically modern humans across Europe that occurred roughly 40,000 years ago. In the last two decades, our view of Neanderthals has changed radically, but at the time of discovery, the idea of the existence of music in Neanderthal culture still seemed revolutionary.

Figure 1. The perforated femur diaphysis no. 652 from Divje babe I with two complete (nos. 2 and 3) and two partially preserved holes (nos. 1 and 4). Soon after discovery, the question arose whether it was a Neanderthal musical instrument or simply a bone pierced and gnawed by a carnivore (photo Tomaž Lauko, NMS).

2. Contestable Explanation of the Carnivore Origin of the Holes

The explanation of the find as a pseudo-artefact was immediately unilaterally taken over by F. d'Errico and colleagues [12], G. Albrecht and colleagues [13], P. G. Chase with A. Nowell [14], and later some others [15,16]. Thus, they negated the potential multilateral significance the find could have had for archaeology and other sciences. Advocates of the carnivore origin of the holes have not rested in the years since the discovery of specimen no. 652. They published a series of articles on the same topic. Among them, d'Errico was the only one who micro-scoped the find and explained the findings of the microscopy in accordance with his previous estimate [12], published in Antiquity in 1998 [17–19]. I. Turk with colleagues [10,20–25] (see also [26]) continuously argumentatively claimed that some of their statements, regarding their explanations about the origin of the holes and damages on the perforated bone, are incorrect [13,14,16,27–29]. To obtain more accurate explanation of the find, I. Turk and colleagues performed and published a series of experiments on perforating fresh brown bear femur diaphyses, using models of wolf, hyena, and bear dentitions (Figure 2), as well as replicas of Palaeolithic tools that were present in various Mousterian levels in DB [20,21,30,31]. Various musical tests of the find were also performed, which was reconstructed several times for this purpose [7,32–37].

After I. Turk and colleagues contested the arguments for the carnivore origin of the holes in numerous publications and offered arguments for their anthropic origin, it was up to advocates of the carnivore origin to refute their findings argumentatively, which they have not done so far. Their discussion of the find is distinctly one-sided and, with one sole exception [13], included no experiments. They presented certain erroneous claims to support their explanation, e.g., about the number of holes [14,19,27], contra [20,22,23], how the holes cannot be made in any other way than by drilling [13,28], contra [10,21,30], the placement of holes on the thinnest parts of the cortical bone [13,14,16], contra [22–24], actual possibilities of teeth grip in connection to holes and gnawing marks [13,14,16,18,19], contra [20,24,25], the sound capabilities of the musical instrument, if that is what the find actually is [19,27], contra [7,36–38], the inappropriateness of a cave bear femur as a support for a musical instrument in comparison to the supports from bird bones [29], contra [7,36–38], and about the frequency of gnawing marks [18] (Figure 9 from Reference [18]), Ref. [19], which in certain cases can also be explained as corrosion formations [10,39]. Corrosion was found to be especially strong in the layer containing the find [10,40].

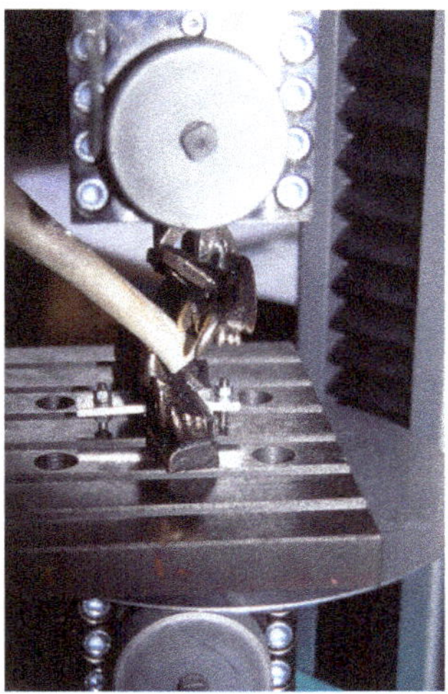

Figure 2. Experimental piercing of a fresh femur of a young brown bear using a bronze model of hyena's dentition and the ZWICK/Z 050 machine for measuring compressive force (photo Ivan Turk, ZRC SAZU).

Supporters of the anthropic origin of the holes were also mistaken; e.g., about the original number of holes [4] and the original length of the musical instrument [35]. The first reconstructions of the find intended to research its musical capabilities, which places the mouthpiece into the large notch on the distal metaphysis, and which consequentially did not consider the opposite hole (at the time supposed to be a thumb hole because of its proximity to the mouthpiece), were also erroneous [31,32,34]. Due to the wrong orientation, the capability of the find as a musical instrument was reduced, and a remnant of the straight edge sharpened from both sides on the proximal part of the diaphysis, which functions on the musical instrument as the cutting edge of the mouthpiece, was overlooked [10,37] (Figure 9 from Reference [10]). It should be noted that we are dealing here with the first example of a bevelled mouthpiece edge. A bevelled mouthpiece edge, which enables better musical performance of the instrument is not known in later Upper Palaeolithic wind instruments, which are made of mammal limb bones. At already thin bone cortex of bird bones, the additional sharpening of the mouthpiece edge is not necessary to achieve better sonority.

When defining the holes on the femur diaphysis no. 652, which are the key component of all wind instruments, we have to start from certain findings of research of all cave bear finds, acquired with wet sieving of all sediments during the excavations of I. Turk, as well as from the findings of his fresh bone piercing experiments. In DB, the main damage to the bones was, in addition to humans, made by wolves (all remains belong to 30 individuals at the most) and not cave hyenas (zero specimens and no indirect proof, such as coprolites and digested bones) [25], contra [16]. The complete and partial holes on the femur diaphysis are undoubtedly of mechanical origin. Namely, both have a funnel-shaped inner edge, which occurs during piercing with a tooth or a pointed tool. Experiments show that the compression of the diaphysis with sharp (unworn) teeth or striking it with a pointed tool result in the longitudinal cracking of the compact bone [20]. Longitudinal cracks are present on some of the fossil bones that were undoubtedly pierced by carnivores [16] (Figures 5 and 6 from Reference [16]). Thus, the femur or some other tubular bone, with removed meta- and epiphyses, usually breaks in half longitudinally during piercing and widening of the hole(s) [16] (Figure 6 from Reference [16]). This is, however, not true for compression and piercing with strongly worn teeth and blunt pointed tools. A crack on the posterior side of the femur diaphysis no. 652 (Figure 1), which zigzags longitudinally from one end to the other is only superficial, and occurs during weathering in the course of fossilization.

It is significantly different from the continuous, rectilinear in-depth crack that occurred on fresh bones during experimental piercing with metal models of carnivore dentition. Since the femur diaphysis no. 652 is not cracked in this way, solely worn teeth or blunt pointed tools can be considered to have produced the holes.

Both partial holes, which advocates of the carnivore origin of the holes considered to be evidence of bites, can be explained differently. V-shaped fractures start on both ends of the diaphysis in the partial hole, meaning that the holes came first and both fractures followed (Figure 1). If the fractures had been made simultaneously with the holes, three cracks would certainly have occurred: Two connected to the fracture and the third one on the diaphysis, with its starting point in the remains of the hole [13] (Figure 10.3 and p. 8, point 4 from Reference [13]). There is no third crack on either of the partial holes. Among 550 cave bear femur diaphyses without epiphyses, similar in size to specimen no. 652 from various layers in DB, only two are pierced and none with the V-fracture and a partial hole.

Judging from the shape and size of the holes, we agree with F. d'Errico [12,18] that they could have been pierced primarily with canines (Figure 3). C. Diedrich [16] believes that all holes in the bones of cave bear from different sites were made exclusively by premolars and molars. According to the first explanation, primarily an adult cave bear is possible, while, according to the second, it would have to be an adult cave hyena which was, like all hyenas, specialised for crushing bones. Frequent in vivo damage on the canine teeth of adult cave bears indicates their rough use. Measured forces from our experiments with models of various carnivore dentitions reveal that piercing with canine teeth takes one-time greater force than piercing with molars and two-times greater force if the tip of canine tooth is blunt [20]. Such forces are on the verge of the capability of the largest carnivores [41,42]. The oval shape of one of the holes and possible antagonist canine impression on the opposite, anterior side connected to it are not in line with the grip and occlusion of canine teeth [10,24], contra [18]. Congruity with the occlusion can be achieved only if the diaphysis is placed lengthwise to the teeth line in the sagittal direction. Such a bite would be highly unlikely, if possible at all. Due to the different shape of teeth tips and shape of the holes (Figure 3) and the unusual longitudinal femur grip considering the only possible dent (pitting after d'Errico [18] (Figure 9 from Reference [18]) of the antagonist tooth [25], cave hyena and the grip with so-called crushing teeth, which is referred to by C. Diedrich [16], is not an option. As stated above, there is also no direct and indirect evidence of the presence of hyena at DB. The same as for the bite of a hyena is true for the bite of a wolf, which is the second best represented carnivore at the site, next to cave bear. The latter is represented with several thousand individuals. It is also not possible to make a partial hole and a complete hole one beneath the other and simultaneously an emphasised depression right by hole no. 3 (Figure 3f, Figure 5) with just any tooth [24], contra [16,18].

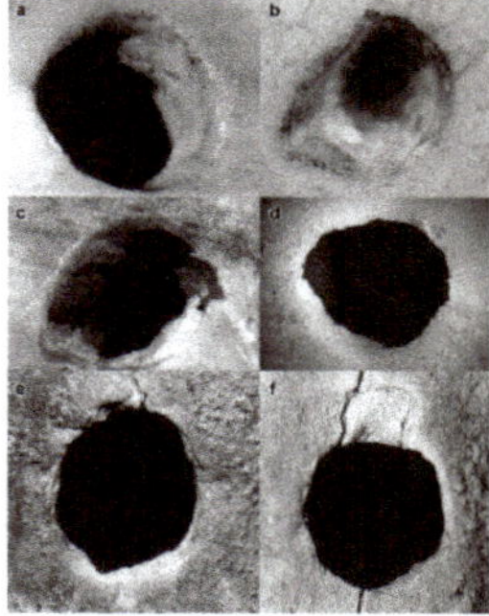

Figure 3. Experimental holes on juvenile femur diaphysis of brown bear made by: (**a**) a bear's canine tooth, hole size 8.2 × 8.2 mm; (**b**) a hyena's lower canine tooth, hole size 6.5 × 8.3 mm; (**c**) a hyena's 3rd upper premolar, hole size 6.5 × 9.0 mm; (**d**) a hole made by a pointed stone tool and bone punch, size 6.0 × 7.4 mm (**e,f**) complete holes no. 2 (size 8.2 × 9.7 mm) and 3 (size 8.7 × 9.0 mm) on the femur from DB no. 652 (ZRC SAZU, Archive of Institute of Archaeology).

Many juvenile femur diaphyses, and other tubular bones of extremities in DB and elsewhere have a bigger distal or proximal semi-circular notch, which is typical carnivore damage. Such a notch also occurs on the distal metaphysis of femur no. 652 from DB (Figure 1). Considering the circumstances, it can be attributed to a wolf, with which P.G. Chase and A. Nowel also agree [14]. Undisputable traces of gnawing on both ends of the diaphysis cannot be linked with certainty to the occurrence of both complete holes and at least one partial hole [10,24], contra [18]. Since it was possible for carnivores to damage Palaeolithic osseous artefacts and leave traces of teeth on them, which is proven by some of the gnawed osseous points [20] (Figure 20 from Reference [20]), ([43] p. 257, Photo 1), this could have happened to femur diaphysis no. 652 at some later time. Most probably, it was at that time that both V-fractures with the starting point in the hole, from which only a partial hole could have remained both times, could have been made.

3. Anthropic Origin of the Holes

Due to the shortcomings the explanation of F. d'Errico and his colleagues regarding the carnivore origin of the holes and damage on femur diaphysis no. 652, more attention is warranted to the alternative explanation of the find and findings connected to it, which are based on the results of appropriate experiments and on indirect evidence from archaeological finds in Mousterian levels of DB.

When piercing bones Neanderthals could imitate carnivores and use pointed tools and the dynamic force of strikes, instead of the compression force of teeth. Holes can be carved into the diaphysis with pointed stone tools [30] found in the Mousterian levels of DB [44]. The bone does not crack during this procedure. The edge of such holes is irregular and serrated, just as with holes on the specimen no. 652, while the edge of holes made by a tooth is generally smooth, depending on the thickness of the cortical bone (Figure 3). Clearly recognisable tool marks are not always present as was attested by F. d'Errico. Namely, six experimentally carved holes were put under microscopic examination. Tool marks were detected on only half of them [19]. However, characteristic damage, such as a broken tip and other fractures, does occur on the tools. Such damage is also present on some of the Mousterian tools from DB [10,20,31] (Figure 4). Holes can also be made with a blunt ad hoc bone punch, struck with a wooden hammer, if a dent has previously been carved into the cortical bone. The holes produced by this technique are morphologically identical to the holes on the specimen no. 652 and completely lack the conventional manufacture marks [21].

Figure 4. Tools suitable for perforating cortical bone: Pointed stone tools (the first on the right has a broken tip) and bone punches from the Mousterian layers of Divje babe I (photo Tomaž Lauko, NMS).

Whether the bone will crack depends on the shape of the punch point (blunt or sharp). In Mousterian levels of DB, beside rare undisputable fragments of bone and antler points, some ad hoc punches with rounded tips were found [23,45] (Figure 4).

At first glance, such artificially made holes on the diaphysis resemble holes made with teeth. The latter are almost always in the vicinity of the epiphyses and only exceptionally on juvenile diaphyses of the approximately same size, such as specimen no. 652 [16]. This is conditioned with the ability of large carnivores, i.e., physical restriction regarding the grip and muscle strength, and with the thinner cortical bone near epiphyses. Unlike animals, man was able to pierce holes along the entire femur diaphysis, regardless of the thickness of the cortical bone. While puncturing bones, people could choose among significantly more methods than animals, which instinctively always do exactly the same. Therefore, in the case of the artefact, it is easier to substantiate the problematic damage, including the above-mentioned depression near hole no. 3 on the posterior side of the diaphysis. Namely, in its vicinity, there are two parallel micro-scores on the abraded surface of the cortical bone (Figure 5), which could be interpreted as cut marks made by stone tools. These micro-scores are never mentioned by advocates of the carnivore origin of the holes. The possibility that people used a femur, the distal end of which was previously damaged by carnivores, is not ruled out. Regarding the absence of other microscopic traces related to manufacture, they could have been erased due to extremely strong corrosion in the layer comprising the find. Only the more distinct scores were preserved, as well as the dent(s) (pitting after F. d'Errico [18]) made by teeth, which, considering their position, cannot be connected with certainty to the production of holes by compression and piercing with teeth. Due to their orientation and shape, all scores and dents, recognized by d'Errico and colleagues, cannot be attributed to carnivores. Carnivores make scores with their teeth that are perpendicular or slightly oblique to the axis of the diaphysis. They are not able to make a score subparallel to the axis of the diaphysis with their bites [10] (Figure 9 from Reference [10]). Some of the dents must have been made by corrosion, which was not considered by d'Errico and colleagues [39]. At least one longitudinal score could be a tool mark.

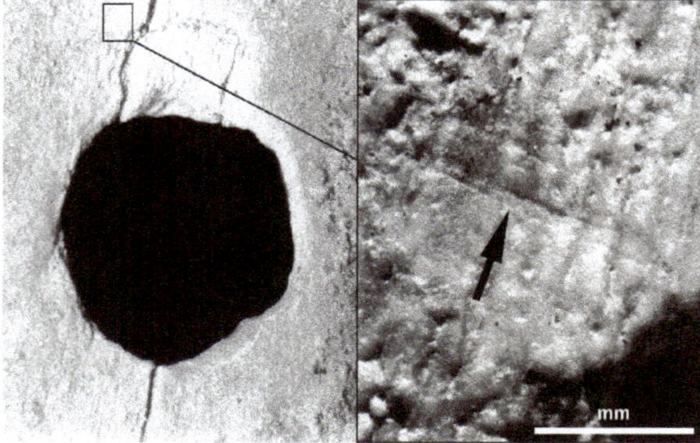

Figure 5. Depression near hole no. 3 on the posterior side of the diaphysis and location of two parallel micro-scores on the abraded surface of cortical bone (marked with an arrow) (ZRC SAZU, Archive of Institute of Archaeology).

The strongest argument for the thesis that, the DB perforated femur is indeed a deliberately crafted musical instrument, comes from experimental musical research on a reconstructed find. It was determined that the artificially transformed juvenile diaphysis is ideal in shape and length for the performance of music using a special playing technique [36,37]. Following the directions of I. Turk [22], in 2010, the missing parts, and both partial holes of the original, were reconstructed on the left cave bear juvenile femur of the size of the original (Figure 6). Due to practical reasons, the mouthpiece of the reconstructed musical instrument was made on the straight edge of the widened part of medullary cavity. This edge fits lips better than the edge of the narrowed part. Later, professional musician L. Dimkaroski established that on the original, the remnant of the straight part of this edge is bevelled on both sides of the cortical bone and could as such function as the perfected cutting edge of the mouthpiece [37]. Considering the position of the edge of the mouthpiece and torsion of the diaphysis, the diaphysis of the left femur is also handier for a right-handed musician, while a right femur diaphysis would be more suitable for a left-handed player. All contemporary music genres can be played on the thus reconstructed musical instrument. The comparative acoustic analysis and tests revealed its great musical capability. With a musical capability of $3\frac{1}{2}$ octaves [37] (a CD in the appendix), the reconstructed musical instrument from Divje babe I surpasses the musical capability of reconstructed Aurignacian osseous wind instruments, made from the bones of large birds [38,46,47].

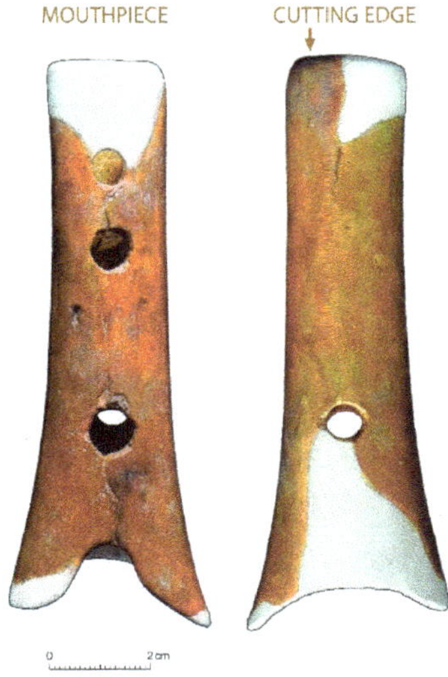

Figure 6. Reconstruction of the Neanderthal musical instrument from Divje babe I. The reconstructed parts are in white plaster. The position of the bevelled cutting edge of the mouthpiece is marked by an arrow (photo Tomaž Lauko, NMS).

4. Conclusions

If the holes on femur no. 652 are not equated with the obvious and frequent impressions of teeth, i.e., punctures with the impressed cortical bone [16] (Figure 4: 9b–11b; Figure 7: 2b from Reference [16]) – on meta- and epi-physes from cave bear sites, as is done by d'Errico and some of his adherents, the find does not have a suitable comparison in collections of pierced limb bones of cave bear [16,48]. The exception is the diaphysis of a juvenile femur with three holes from the Aurignacian cave site Istállóskő in Hungary [6,49], which is currently not considered to be a potential musical instrument, due to numerous more convincing new finds of Aurignacian wind instruments in cave sites of the Swabian Jura [50] and the French Pyrenees [51].

Currently, this unique find fulfils all conditions on the basis that it can be defined as the oldest known musical instrument. These are: clear archaeological and stratigraphic context [44], dating [8,9,11], explanation of manufacturing [21], musical verification [36,37], ([47] (p. 458), contra [19] p. 55), and good comparisons in later periods [52]. In a preserved state, the find is not suitable for playing music. Playing was enabled by the reconstruction based on concrete data and the well-founded assumption that the reconstructed parts were removed by a wolf, prior to cementation. Similarly damaged is the Upper Palaeolithic musical instrument from the loess layer in the open-air site Grubgraben (Austria) [52]

(Figures 2 and 3 form the Reference [52]). Due to the fine texture of the loess, the damage on the Grubgraben flute could have occurred exclusively prior to its inclusion into the sediment. Presumably, it was damaged by a wolf occasionally feeding on the remains of the prey of Palaeolithic hunters.

The find of the Mousterian musical instrument from DB has certain advantages in relation to other, declaratively the oldest similar instruments from the sites in the Swabian Jura, in regards to context, stratigraphy, reconstruction, and morphometric characteristics. The context and stratigraphy, supported by indirect ^{14}C AMS, U/TH, and ESR dates [8,9,11], are indisputable because the find, firmly cemented into the breccia, could not move within the sediment or be mixed with older finds due to detected gaps in sedimentation. One of them occurred above the cemented part of layer 8, where the musical instrument was found [53].The leading Aurignacian artefact – the point with the split base – found in the youngest Pleistocene combined layer 2–3, two metres higher, enables the cultural paralleling of Aurignacian level with sites of the Swabian Jura and simultaneously indisputably proves the greater relative age of specimen no. 652, in comparison to the finds of musical instruments in the Swabian Jura and elsewhere [50].

The reconstruction of the DB musical instrument is more reliable than the reconstructions of Swabian finds, in which either the length or the mouthpiece are not preserved [46,47,50,54]. The morphometric characteristics of the DB musical instrument are such that, even on the basis of physical laws, despite the smaller length, they enable a greater musical capability in comparison to Swabian finds. Excuses of everything being dependent on the interpreter are only partly valid. This has also been confirmed by the unpublished musical experiments of professional musician L. Dimkaroski (personal communication) with the replica of Swabian wind instrument GK 1 and comparative acoustic calculations of F. Z. Horusitzky [38] for Swabian instruments GK 1 and GK 3.

The musical instrument from Divje babe I, which predates 50 ka, firmly supported with a Mousterian (Neanderthal) context and chronology, remains the strongest material evidence so far for Neanderthal musical behaviour. According to the present knowledge and archaeological context of the find, there are no obstacles for the find to be interpreted as a Neanderthal musical instrument.

Author Contributions: Conceptualization, M.T. and I.T.; methodology, I.T.; investigation, I.T. and M.T.; writing—original draft preparation, M.T., I.T. and M.O.; writing—review and editing, M.T.; supervision, M.O. All authors have read and agreed to the published version of the manuscript.

Funding: The authors acknowledge the financial support from the Slovenian Research Agency (P6-0283 and P6-0064). We also thank ZRC SAZU Institute of Archaeology for part-financing from its current assets.

Conflicts of Interest: The authors declare no conflict of interest.

References

1. Turk, I.; Dirjec, J.; Kavur, B. The oldest musical instrument in Europe discovered in Slovenia? *Razpr. 4. Razreda Sazu* **1995**, *36*, 287–293.

2. Turk, I.; Dirjec, J.; Kavur, B. Description and explanation of the origin of the suspected bone flute. In *Mousterian «Bone Flute» and Other Finds from Divje Babe I Cave Site, Slovenia*; Turk, I., Ed.; Opera Instituti Archaeologici Sloveniae 2: Ljubljana, Slovenia, 1997; pp. 157–175.

3. Lau, B.; Blackwell, B.A.B.; Schwarcz, H.P.; Turk, I.; Blickstein, J.I. Dating a flautist? Using ESR (Electron spin resonance) in the Mousterian cave deposits at Divje babe I, Slovenia. *Geoarchaeology* **1997**, *12*, 507–536. [CrossRef]

4. Otte, M. On the suggested bone flute from Slovenia. *Curr. Anthropol.* **2000**, *41*, 271–272. [CrossRef] [PubMed]

5. Otte, M. La mobilité rapide, caractère propre au Paléolithique supérieur d'Eurasie. In *Modes de Contacts et de Déplacements au Paléolithique Euroasiatique/Modes of Contact and Mobility during the Eurasian Palaeolithic*; Otte, M., Ed.; 28-31 mai 2012. ERAUL 140; Actes du Colloque International de la Commission 8 de l'UISPP, Université de Liège: Liège, Belgium, 2014; pp. 693–706.

6. Horusitzky, F.Z. Les flûtes paléolithiques: Divje Babe I, Istállóskő, Lokve, etc. Point de vue des experts et des contestataires (Critique de l'appréciation archéologique du spécimen no. 652 de Divje Babe I, et arguments pour la défence des spécimens Pb 51/20 et Pb 606 de MNM de Budapest). *Arheol. Vestn.* **2003**, *54*, 45–66.

7. Horusitzky, F.Z. Analyse acoustique de la flûte avec souffle proximal. In *Divje babe I. In Upper Pleistocene Palaeolithic Site in Slovenia. Part 2: Archaeology*; Turk, I., Ed.; Opera Instituti Archaeologici Sloveniae 29: Ljubljana, Slovenia, 2014; pp. 223–233.

8. Blackwell, B.A.B.; Yu, E.S.K.; Skinner, A.R.; Turk, I.; Blickstein, J.I.B.; Turk, J.; Yin, V.S.W.; Lau, B. ESR-Dating at Divje babe I, Slovenia. In *Divje Babe I: Upper Pleistocene Palaeolithic site in Slovenia, Part 1: Geology and Palaeontology*; Turk, I., Ed.; Opera Instituti Archaeologici Sloveniae 13: Ljubljana, Slovenia, 2007; pp. 123–157.

9. Blackwell, B.A.B.; Yu, E.S.K.; Skinner, A.R.; Turk, I.; Blickstein, J.I.B.; Skaberne, D.; Turk, J.; Lau, B. Dating and paleoenvironmental interpretation of the Late Pleistocene archaeological deposits at Divje Babe I, Slovenia. In *The Mediterranean from 50 000 to 25 000 BP: Turning Points and New Directions*; Calbet, M., Szmidt, C., Eds.; Oxbow Books: Oxford, UK, 2009; pp. 179–210.

10. Turk, M.; Turk, I.; Dimkaroski, L.; Blackwell, B.A.B.; Horusitzky, F.Z.; Otte, M.; Bastiani, G.; Korat, L. The Mousterian musical instrument from the Divje babe I cave (Slovenia): Arguments on the material evidence for Neanderthal musical behaviour. *L'anthropologie* **2018**, *122*, 679–706. [CrossRef]

11. Nelson, D.E. Radiocarbon dating of bone and charcoal from Divje babe I cave. In *Mousterian «Bone Flute» and Other Finds from Divje Babe I Cave site, Slovenia*; Turk, I.. Ed.; Opera Instituti Archaeologici Sloveniae 2: Ljubljana, Slovenia, 1997; pp. 51–64.

12. d'Errico, F.; Villa, P.; Pinto Llona, A.C.P.; Idarraga, R.R.A. Middle Palaeolithic origin of music? Using cave-bear bone accumulations to assess the Divje Babe I bone "flute". *Antiquity* **1998**, *72*, 65–79.

13. Albrecht, G.; Holdermann, C.S.; Kerig, T.; Lechterbeck, J.; Serangeli, J. "Flöten" aus Bärenknochen—Die frühesten Musikinstrumente? *Archäologisches Korresp.* **1998**, *28*, 1–19.

14. Chase, P.G.; Nowell, A. Taphonomy of a suggested Middle Paleolithic bone flute from Slovenia. *Curr. Anthropol.* **1998**, *39*, 549–553. [CrossRef]

15. Morley, I. *The Prehistory of Music*; University Press: Oxford, UK, 2013; pp. 1–280.

16. Diedrich, C.G. 'Neanderthal bone flutes': Simply products of Ice Age spotted hyena scavenging activities on cave bear cubs in European cave bear dens. *R. Soc. Open Sci.* **2015**, *2*, 140022. [CrossRef]

17. d'Errico, F. Just a bone or flute? The contribution of taphonomy and microscopy to the identification of prehistoric pseudo-musical instruments. In *The Archaeology of Sound: Origin and Organisation*; Hickmann, E., Kilmer, A.D., Eichmann, R., Eds.; Studien zur Musikarchäologie, 3, Orient-Archäologie, 10: Rahden/Westf., Germany, 2002; pp. 89–90.

18. d'Errico, F.; Henshilwood, C.; Lawson, G.; Vanhaeren, M.; Tillier, A.M.; Soressi, M.; Bresson, F.; Maureille, B.; Nowell, A.; Lakarra, J.; et al. Archaeological evidence for the emergence of language, symbolism, and music: An alternative multidisciplinary perspective. *J. World Prehistory* **2003**, *17*, 1–70.

19. d'Errico, F.; Lawson, G. The sound paradox: How to assess the acoustic significance of archaeological evidence? In *Archaeoacoustics*; Lawson, G., Scarre, C., Eds.; McDonald Institute Monographs: Cambridge, UK, 2006; pp. 41–57.

20. Turk, I.; Dirjec, J.; Bastiani, G.; Pflaum, M.; Lauko, T.; Cimerman, F.; Kosel, F.; Grum, J.; Cevc, P. New analyses of the »flute« from Divje babe I (Slovenia). *Arheol. Vestn.* **2001**, *52*, 25–79.

21. Turk, I.; Bastiani, G.; Blackwell, B.A.B.; Horusitzky, F.Z. Putative Mousterian flute from Divje babe I (Slovenia): Pseudoartefact or true flute, or who made the holes? *Arheol. Vestn.* **2003**, *54*, 67–72.

22. Turk, I.; Pflaum, M.; Pekarovič, D. Results of computer tomography of the oldest suspected flute from Divje babe I (Slovenia): Contribution to the theory of making holes in bones. *Arheol. Vestn.* **2005**, *56*, 9–36.

23. Turk, I.; Blackwell, B.A.B.; Turk, J.; Pflaum, M. Results of computer tomography of the oldest suspected flute from Divje babe I (Slovenia) and its chronological position within global palaeoclimatic and palaeoenvironmental change during Last Glacial. *L'Anthropologie* **2006**, *110*, 293–317. [CrossRef]

24. Turk, I.; Dirjec, J.; Turk, M. Flute (musical instrument) 19 years after its discovery. Critique of the taphonomic interpretation of the find. In *Divje babe I. Upper Pleistocene palaeolithic site in Slovenia. Part 2: Archaeology*; Turk, I., Ed.; Opera Instituti Archaeologici Sloveniae 29: Ljubljana, Slovenia, 2014; pp. 235–268.

25. Turk, I.; Turk, M.; Toškan, B. Could a cave hyena have made a musical instrument? A reply to Cajus G. Diedrich. *Arheol. Vestn.* **2016**, *67*, 401–407.

26. Nițu, E.C. Some observations on the supposed natural origin of the Divje babe I flute. *Ann. D'université Valahia Targovistesection D'archéologie Et D'histoire* **2015**, *17*, 33–44.

27. Nowell, A.; Chase, P.G. Is a cave bear bone from Divje Babe, Slovenia, a Neanderthal flute? In *The Archaeology of Early Sound: Origin and Organization*; Hickmann, E., Eichmann, R., Eds.; Studien zur Musikarchäologie, 4, Orient-Archäologie, 12: Rahden/Westf., Germany, 2002; pp. 69–81.

28. Albrecht, G.; Holdermann, C.S.; Serangeli, J. Towards an archaeological appraisal of specimen N° 652 from Middle-Palaeolithic level D/(Layer 8) of the Divje babe I. *Arheol. Vestn.* **2001**, *52*, 11–15.

29. Holdermann, C.S.; Serangeli, J. Einige Bemerkungen zur Flöte von Divje babe I (Slowenien) und deren Vergleichsfunde aus dem österreichischen Raum und angrenzenden Gebieten. *Archäologie Österreichs* **1998**, *9*, 31–38.

30. Bastiani, G.; Turk, I. Results from the experimental manufacture of a bone flute with stone tools. In *Mousterian "Bone Flute" and Other Finds from Divje Babe I Cave Site in Slovenia*; Turk, I., Ed.; Opera Instituti Archaeologici Sloveniae 2: Ljubljana, Slovenia, 1997; pp. 176–178.

31. Kunej, D.; Turk, I. New perspectives on the beginning of music: Archaeological and musicological analysis of a Middle Palaeolithic Bone «Flute». In *The Origins of Music*; Wallin, N.L., Merker, B., Brown, S., Eds.; MIT Press: Cambridge, MA, USA; London, UK, 2000; pp. 235–268.

32. Omerzel-Terlep, M. Bone flutes. The beginning of the history of instrumental music in Slovenia, Europe and the world. *Etnolog* **1996**, *6*, 292–294.

33. Omerzel-Terlep, M. A typology of bone whistles, pipes and flutes and presumed palaeolithic wind instruments in Slovenia. In *Mousterian «Bone Flute» and Other Finds from Divje Babe I Cave Site, Slovenia*; Turk, I., Ed.; Opera Instituti Archaeologici Sloveniae 2: Ljubljana, Slovenia, 1997; pp. 199–218.

34. Kunej, D. Acoustic findings on the basis of the reconstruction of a presumed bone flute. In *Mousterian «Bone Flute» and Other Finds from Divje Babe I Cave Site, Slovenia*; Turk, I., Ed.; Opera Instituti Archaeologici Sloveniae 2: Ljubljana, Slovenia, 1997; pp. 185–197.

35. Fink, B. Neanderthal Flute. Oldest Musical Instrument's 4 Notes Matches 4 of do, re, Mi scale. Musicological Analysis. 1997. Available online: https://web.archive.org/web/20070219094811/http://www.greenwych.ca:80/fl-compl.htm (accessed on 3 February 2020).

36. Dimkaroski, L. Musikinstrument der Neanderthaler. Zur Diskussion um die moustérienzaitliche Knochenflöte aus Divje babe I, Slowenien, aus technischer und musikologischer Sicht. *Mittelungen Der Berl. Ges. Für Anthropol. Ethnol. Und Urgescichte* **2011**, *32*, 45–54.

37. Dimkaroski, L. Musical research into the flute. From suspected to contemporary musical instrument. In *Divje Babe I. Upper Pleistocene Palaeolithic site in Slovenia. Part 2: Archaeology*; Turk, I., Ed.; Opera Instituti Archaeologici Sloveniae 29: Ljubljana, Slovenia, 2014; pp. 215–222.

38. Horusitzky, F.Z. *Comparaison Musicale De L'instrument De Divje Babe Avec Les Trouvailles de la Montagne De Souabe*; Archive of ZRC SAZU Institute of Archaeology: Ljubljana, Slovenia, 2017; pp. 1–11, Unpublished.

39. Tuniz, C.; Bernardini, F.; Turk, I.; Dimkaroski, L.; Mancini, L.; Dreossi, D. Did Neanderthals play music? X-ray computed microtomography of the Divje babe 'flute'. *Archaeometry* **2012**, *54*, 581–590. [CrossRef]

40. Turk, I.; Skaberne, D.; Blackwell, B.A.B.; Dirjec, J. Assessing humidity in the Upper Pleistocene karst environment. Palaeomicroenvironments at Divje babe I, Slovenia. In *Water and Life in a Rocky Landscape. Kras*; Mihevc, A., Ed.; ZRC Publishing: Ljubljana, Slovenia, 2005; pp. 173–198.

41. Christiansen, P.; Adolfssen, J.S. Bite forces, canine strength and skull allometry in carnivores (Mammalia, Carnivora). *J. Zool.* **2005**, *266*, 133–151. [CrossRef]

42. Christiansen, P.; Wroe, S. Bite forces and evolutionary adaptations to feeding ecology of carnivores. *Ecology* **2007**, *88*, 347–358. [CrossRef]

43. López Bayón, I.; Straus, L.G.; Léotard, J.-M.; Lacroix, P.; Teheux, E. L'industrie osseuse du Magdalenien du Bois Laiterie. In *La grotte du Bois Laiterie*; Otte, M., Straus, L.G., Eds.; ERAUL 80: Liège, Belgium, 1997; pp. 257–277.

44. Turk, I. (Ed.) *Divje Babe I. Upper Pleistocene Palaeolithic site in Slovenia. Part 2: Archaeology*; Opera Instituti Archaeologici Sloveniae 29: Ljubljana, Slovenia, 2014; pp. 1–456.

45. Turk, M.; Košir, A. Mousterian osseous artefacts? The case of Divje babe I, Slovenia. *Quat. Int.* **2016**, *450*, 103–115. [CrossRef]

46. Potengowski, A.F.; Münzel, S.C. Die musikalische »Vermessung« paläolithischeer Blasinstrumente der Schwäbischen Alb anhand von Rekonstruktionen. Anblastechniken, Tonmaterial und Klangwelt. *Mitt. Der Ges. Für Urgesch.* **2015**, *24*, 173–191.

47. Conard, N.J.; Malina, M.; Münzel, S.C. Eine Mammutelfenbeinflöte aus dem Aurignacien des Geissenklösterle. Neue Belege für eine musikalische Tradition im frühen Jungpaläolithikum auf der Shwäbischen Alb. *Archäologisches Korresp.* **2004**, *34*, 447–462.

48. Brodar, M. Fossile Knochendurchlochungen. In *Ivan Rakovec Volume*; Grafenauer, S., Pleničar, M., Drobne, K., Eds.; Razprave 4. Razreda SAZU: Ljubljana, Slovenia, 1985; Volume 26, pp. 29–48.

49. Horusitzky, F.Z. Eine Knochenflöte aus der Höhle von Istállóskő. *Acta Archeol. Acad. Sci. Hung.* **1955**, *5*, 133–140.

50. Conard, N.J.; Malina, M.; Münzel, S.C. New flutes document the earliest musical tradition in southwestern Germany. *Nature* **2009**, *460*, 737–740. [CrossRef]

51. Buisson, D. Les flûtes paléolithiques d'Isturitz (Pyrénées-Atlantiques). *Bull. De La Société Préhistorique Française* **1990**, *87*, 420–433. [CrossRef]

52. Einwögerer, T.; Käfer, B. Eine Jungpaläolitische Knochenflöte aus der Station Grubgraben bei Kammern, Niederösterreich. *Archäologisches Korresp.* **1998**, *28*, 21–30.

53. Skaberne, D.; Turk, I.; Turk, J. The Pleistocene clastic sediments in the Divje babe I cave, Slovenia, Palaeoclimate (part 1). *Palaeogeogr. Palaeoclimatol. Palaeoecol.* **2015**, *438*, 395–407. [CrossRef]

54. Münzel, S.; Seeberger, F.; Hein, W. *The Geissenklösterle Flute—Discovery, Experiments, Reconstruction. Studien zur Musikarchäologie III: Archäologie früher Klangerzeugung und Tonordnung. Musikarchäologie in der Ägäis und Anatolien*; Orient-Archäologie 10: Rahden/West, Germany, 2002; pp. 107–118.

 *applied sciences*

Article

The *Carabattola*—Vibroacoustical Analysis and Intensity of Acoustic Radiation (IAR)

Lamberto Tronchin [1,*], Massimiliano Manfren [2] and Vincenzo Vodola [1]

[1] Department of Architecture, University of Bologna, Via dell'Università 50, 47521 Cesena, Italy; vincenzo.vodola2@unibo.it

[2] Faculty of Engineering and Physical Sciences, University of Southampton, Highfields, Southampton SO17 1BJ, UK; m.manfren@soton.ac.uk

* Correspondence: lamberto.tronchin@unibo.it

Received: 21 August 2019; Accepted: 15 January 2020; Published: 16 January 2020

Abstract: Among the studies of musical instruments, one important, sometime underestimated discipline, is represented by ethnomusicology. The acoustic analyses on ethnic musical instruments (M.I.) are much more infrequent if compared to those on classical M.I. This article deals with the vibro-acoustic analysis on one of the most unknown ethnic, Italian M.I., i.e., the *carabattola* (also called *battola*), which used to be played in Italy until the late 1960s during the Holy Thursday before Easter. The study includes modal analysis and Intensity of Acoustic Radiation measured on an original *carabattola*, which was played in the Romagna area until the early twentieth century. After a brief overview about the theory of acoustic and vibrational analysis on musical instruments, the Intensity of acoustic radiation and its correlation with modal analysis are recalled, based on previous studies. In the experimental part of the article, the measurements conducted on the *carabattola* are described. Afterwards, the results obtained both from modal analysis and IAR measurements are analyzed and compared with other measurements previously conducted on musical (particularly percussion) instruments and commented.

Keywords: sound efficiency; intensity of acoustic radiation (IAR); *Carabattola*; modal analysis

---

1. Introduction

The physics of musical instruments represents one of the most intriguing field of acoustics, especially for those scientists that are normally involved in the preservation of cultural heritage. There are different techniques to study the vibro-acoustical behavior of musical instruments; between them, modal analysis and acoustic radiation are usually used. However, some other techniques have been developed starting from these fundamental methods.

The studies on the physics and acoustics of musical instrument that were carried out in the last 30 years normally regarded classical musical instruments, especially violins [1], piano [2,3], wind instrument [4], and other musical instruments. Only a few studies have analyzed other musical instruments [5]. This was due to the request of knowledge of the sound characteristics of those instruments from lutherie, industrial manufactory companies, researchers, curators, collectors, museums, historians, musicians, theatre companies and all parties involved with preservation and restoration of those important, valuable objects. One more reason for studying the sound characteristics of musical instruments from the physical perspective is the emulation of their sound characteristics [6] by means of measurements of impulse responses [7] and convolution with dry music, including nonlinear properties [8,9].

On the other hand, other musical instruments have been developed or invented for several reasons, sometimes very different from each other, but the scientific community did not pay proper attention to them. The *carabattola*, sometime called *battola*, belongs to this category of musical instruments (M.I.).

In order to determine the acoustic characteristics of M.I., acoustic radiation (like acoustic impedance or admittance) is one of the most important physical parameters utilized to characterize their properties. Sound radiation is closely linked to modal patterns, and a connection should therefore have occurred between resonance frequencies and sound production in vibration constructions. Evidence of the negative correlation between acoustic radiation and Frequency Response Function (FRF) of membranes or plates in instruments, such as piano and harpsichord, has been identified in earlier studies by Suzuki [10] and Giordano [11]. The complexity of pianos and harpsichord was perhaps the reason for the negative correlation, and their complex structure could have impeded their understanding of their sound radiation.

Percussion instruments, on the other side, are comparatively straightforward, and frequency response studies, modal analysis, acoustic radiation and the relationship between FRF and noise radiation can be readily discovered [12]. The comparison between experimental modal patterns and previously outcomes could suggest the most appropriate measurement technique for characterizing vibro-acoustical properties of musical instruments. For tympani, where sound generation and modal analysis are related, a vibro-acoustic parameter was needed, capable of correctly correlating noise output with FRF. Applications of this extend beyond musical acoustics into the modeling of musical instruments in auditoria. The search of the link between sound radiation and modal analysis in musical instruments has been an issue for several years.

2. The *Carabattola*

In western music there are plenty of musical instruments normally played since the beginning of prehistoric ages. Most of them were idiophonic M.I., in which the sound was obtained by hitting the bore with wooden or (later) metallic elements. Moreover, there are several instruments which were invented not for generating pitched music but rather for producing acoustic effects, mainly for specific religious circumstances. Among them, in Mediterranean Europe, several ethnographic instruments have been utilized since the early middle ages almost up to the present. The *carabattola* is one of these M.I.

The *carabattola*, also known as *battola*, is an extremely uncommon idiophone instrument of ethnographic music. Its name recalls the origin of sound (acoustic) emission, i.e., it must be hit by some metallic components in order to produce acoustic effect. This instrument used to be performed only during the Holy Week before Easter, and it may have Byzantine origins. With his handler, the player retains the instrument and rapidly turns the *carabattola* right and left. In this way, the movement causes a clapper to hit in rapid sequence alternatively two metallic little circles, inserted in the wood. Thus, the clapper looks like a knocker and the acoustic effect appears.

The *carabattola* belongs to a series of musical instruments, all of them played during Holy Week before Easter in Mediterranean Europe. Other similar musical instruments are the *Cembalo* or *Crotalo* as Francesco Saverio Quadria, already reported in 1734 [13], normally made in oak or chestnut wood. Some other similar musical instruments, like *Matracula*, (Figure 1) are still used, although seldomly, in some parts of Italy such as the Sardinia region [14].

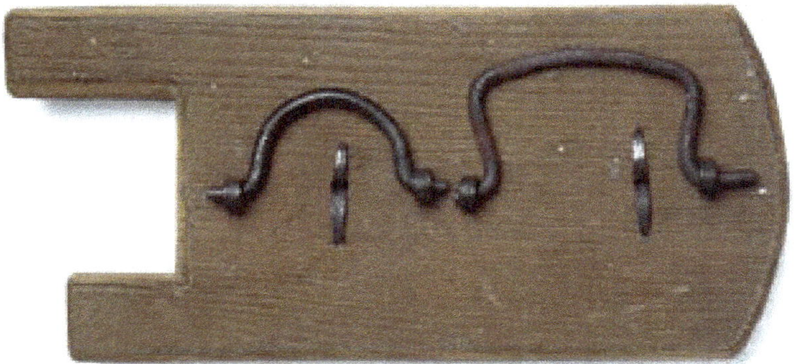

Figure 1. *Matracula*, a percussion instrument still in use in Sardinia (Italy).

The *Carabattola*, as well as the other similar musical instruments specifically realized for Holy Week, does not provide a specific pitched sound, but rather a particular acoustic effect, which increases the particular religious atmosphere, together with the chanting and the incense. These specific characteristics are the reasons for the decision to carry out vibrational and acoustic characterization of the wooden element (the "soundboard").

In this paper, an original example of *Carabattola* is analyzed. This instrument belongs to the Madonna del Carmine Church in Bagnacavallo, Ravenna, Northern Italy. It was locally called *Scarabàtla* and used to be played roughly until 1970. The chest (probably made by wood from fruit trees) is approximately 25 cm width, 50 cm length, 2.2 cm thickness, whilst the metallic (perhaps iron) circles are 2.0 cm width. It should be emphasized that this tool provides a background noise, partly comparable to the noise of a grater and not a real sound (as usually expected).

3. Material and Methods

3.1. Acoustic Radiation

The efficiency of acoustic radiation is a measure of the effectiveness of a vibrating surface in generating sound power. It could be defined by the relationship:

$$\sigma = \frac{W}{\rho_0 c S \left\langle \overline{v_n^2} \right\rangle} \tag{1}$$

in which W is the sound power radiated by a surface with area S, which could be obtained by integrating the far-field intensity over a hemispherical surface centered on the panel, and $\left\langle \overline{v_n^2} \right\rangle$ is the space-averaged value of the time-averaged normal distribution of velocity [15]. This parameter could be utilized for searching a link between movement (vibration) and acoustic generation (sound).

Various measuring techniques that are helpful for noise emission analysis could be acquired from this overall concept, ranging from frequent applications in noise emissions in machinery, to rare examples on musical instruments. In musical acoustics, previous investigations have been carried out on piano and harpsichord soundboards using different technique. Wogram used the parameter F/v, defining F as the excitation force and v as the resulting velocity at the point of excitation [16]. In his experiments, he found a maximum at a frequency that is close to or below 1 kHz and that falls sharply below 100 Hz and above 1 kHz. He discovered that it falls typically by a factor of 10, as the frequency varies between 1 and 5 kHz. On the other hand, the "surface intensity method" was identified by Suzuki [10] and defined by the equation:

$$I = \mathrm{Re}[p(a/j\omega^*)/2] \tag{2}$$

where I represents the average intensity in time, perpendicular to the vibrating surface, measured in near field (about 30 cm from the radiating surface); ω is the angular frequency; Re and * are the real part and the complex conjugate of a complex number; p and a are the pressure and the normal acceleration at the measuring point. Further, Giordano used the parameter p/v where p is the sound pressure measured in near field, and v is the velocity of the soundboard [11]. At around 1 kHz in all sampled frequencies p/v is larger and drops below a few hundred hertz and above 5 kHz.

It is essential to note that all these studies have one prevalent outcome: The frequencies of resonance have not coincided with those of acoustic emission, but they often have an adverse correlation.

3.2. Intensity of Acoustic Radiation—IAR

From the experiment described above, the Intensity of Acoustic Radiation (IAR) parameter was defined in 2005 as the space averaged amplitude of the cross-spectrum between the sound pressure generated from the vibration of the surface and the velocity of the surface vibration [17]:

$$IAR(\omega) := \quad \langle P(\omega)*V(\omega) \rangle \tag{3}$$

For the measurement, an omnidirectional microphone is required. According to Suzuki, the microphone must have been positioned in a specific point which corresponds to about one-quarter of the wavelength of the frequency corresponding to the earliest mode. If tympani are involved, this distance is approx. 25 cm over the membrane.

The measurements should be carried out in a slightly reverberating space where average radiation induced by early modes is easily achieved via the reverberation time. The acoustics chamber has no effect on the readings at higher frequencies. In addition, the measurements are enhanced by space-averaging of information carried out by shifting transductors through the instruments. It should be considered that other vibroacoustic parameters have been obtained starting from IAR, applied in very different field of acoustics, as building acoustics [18].

3.3. I.A.R.: Previous Measurements

When IAR was defined, in order to verify the robustness of the parameter, two distinct percussion instruments were used to measure IAR. These instruments were two kettledrums. The first was a 25-inch (approx. 65 cm) kettledrum plexiglass Adam with a Remo mylar skin and central reinforcement tuned at about 166 Hz (corresponding to E), whilst the second one was a copper 25-inch Ludwig kettledrum, which was tuned to roughly 145 Hz (D-corresponding) with a mylar skin without a central reinforcer.

The tympani measurements were carried out in two different ways. A hammer with an accelerometer was used in the first case, whilst the hammer was replaced by a shaker in the second case. The microphone was in the same position in both cases. In order to verify the outcomes from previous researches, up to 15 modes were obtained and compared with literature. This allowed verifying the accuracy of the measurements. Figure 2 reports the results obtained in 2005 [17].

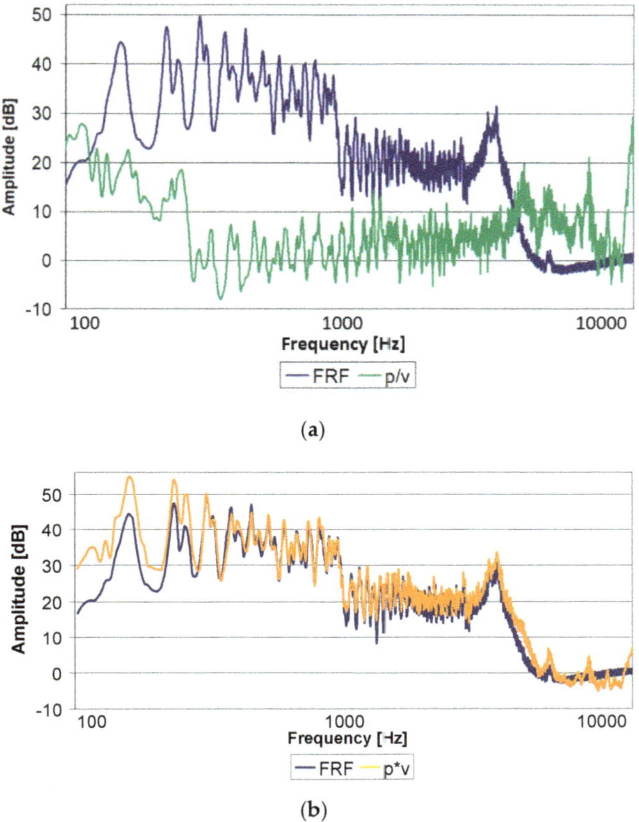

Figure 2. Tympani: (**a**) FRF compared with *p/v*; (**b**) FRF compared with IAR.

4. Measurements on *Carabattola*

The measurements on the *carabattola* were carried out in a similar way as the tympani. In one of two metal circles, the shaker excited the instrument, with the microphone positioned approximately 25 cm above. All measures were carried out in the same room of the tympani, with the same acoustic boundary conditions as described in the article [17].

Modal analyzes were also performed in this case. The measurements were conducted positioning the accelerometers on 24 positions and stored (Figure 3). In a further step, all the measurements were post processed in order to obtain FRF and modal patterns. The findings are quite different from the kettledrum, because the sound generation varies significantly from the tympani. The IAR is measured by the proportion *p/v* and FRF as depicted in Figure 4. Due to the increase in modal density at higher frequencies, the chart is restricted to 1 kHz. The modal patterns from 107 to 344 Hz, measured in the instrument, are reported in Figure 5.

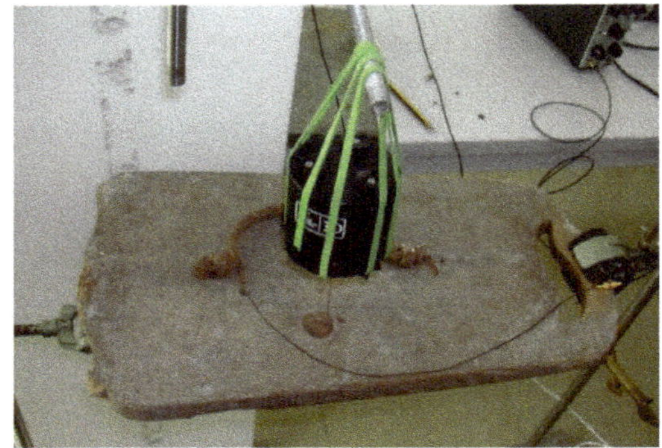

Figure 3. The measurements on the *carabattola*.

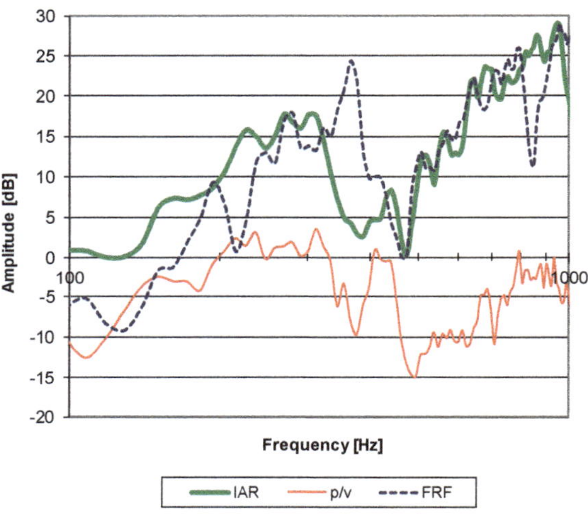

Figure 4. *Carabattola*: FRF compared with *p/v* and IAR.

The match between IAR and FRF in *carabattola* is not as strong as in kettledrums. However, the negative correlation is not so obvious from FRF to *p/v*. The unique properties of sound emission of *carabattola*, which significantly differ from the tympanum, should explain this result.

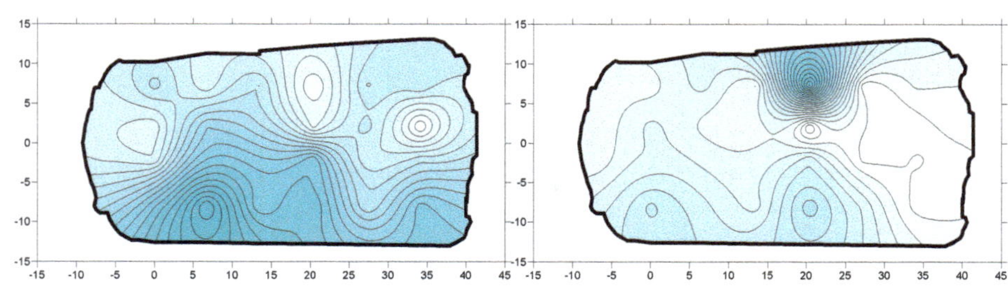

Figure 5. *Cont.*

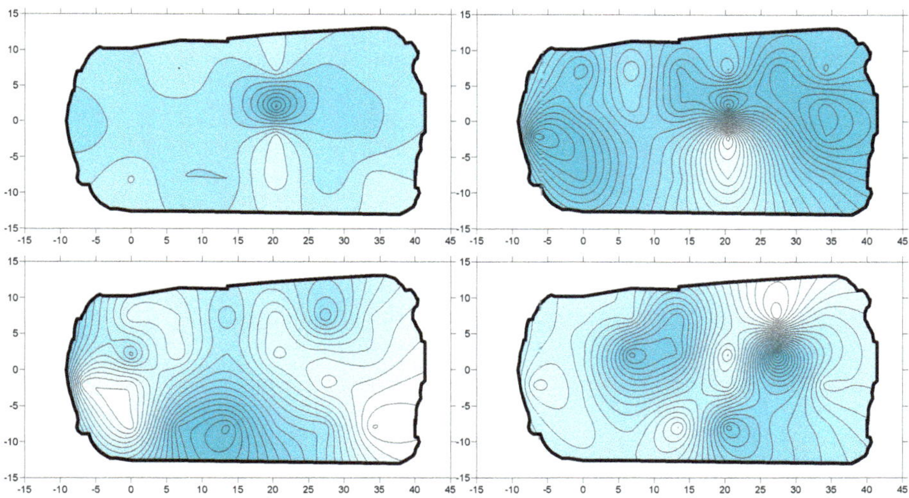

Figure 5. Modal patterns for the *carabattola* (from top left to bottom right): 107 Hz, 150 Hz, 193 Hz, 247 Hz, 297 Hz, 322 Hz.

The IAR measurements on the *carabattola* were also carried out together with modal analysis as previously made for the kettledrums. As mentioned earlier, the sound pressure p, as previously reported from Suzuki, Giordano and Tronchin, was evaluated in the near field at 25 cm from the instruments, maintaining the correct distance already proposed in early studies.

5. Results

Analysis of Measured Data

Comparing the results obtained on the *carabattola* with the results obtained on the kettledrums, the following points could be underlined.

In the case of the tympani, the negative correlations between FRF and IAR are very high, and at the same time, the two respective graphs are almost coincident. It should be noted that tympanum has a clearly defined sound pitch, mainly due to the membrane motion.

On the other hand, different results were carried out in the case of *carabattola*, as the mechanism of sound generation differs considerably from the tympanum. Figure 5 shows the first six modal *carabattola* patterns. It is important to note that the metallic plate where the knocker strikes the instrument affects the motion of the sound chest from the evaluation of these patterns. In view of the relationship between FRF and p/v, only partly a negative connection is noted, particularly at medium-low frequencies. In comparison, they are only partly linked, when comparing the two graphs FRF and IAR. The two charts seem like they are shifted of a small frequency range, even if they are much more linked than FRF and p/v.

The tympanum mainly produces sound from the membrane, which is a very straightforward perception of the pitch caused by the membrane vibration. On the other hand, in the *carabattola* the chest and metal knocker produce a sound that is not perceived as a particular pitch.

In the case of tympani, IAR, FRF and p/v are highly correlated each other, as almost all sound generation comes from the membrane (where the sound velocity v is measured), whilst the correlation in *carabattolas* is not as obvious as the sound generation comes from wood alone.

6. Discussion

Two types of percussion instruments, two tympani and one *carabattola*, were carried out with acoustic radiation measurements and modal analysis, as proposed by many authors in previous articles. In both cases, the instruments were excited by means of a shaker, connected with a thin metallic bar. In a previous paper, the shaker resulted better than the head impedance hummer, as the resonance of the bar that connects the shaker with the instruments was discovered at approximately 3 kHz.

The mappings of individual vibration modes were very clear for all the instruments, and in the case of tympani, the frequency ratios were approximately consistent with the theory. For circular and mixed vibration modes, a high degree of correspondence was obtained, while the diametric modes produced slightly less frequencies than the theoretical ones. Due to the typical noise generated in the *carabattola*, modal pattern mapping was not so evident like in the tympanum.

The tympanum generates its sound mainly from the membrane and has a very little thickness if compared with its surface and could be considered bi-dimensional. The membrane is elastically deformable, and its vibration is only slightly damped. Conversely, the *carabattola* is made of a wooden body having a relevant thickness of 2.2 cm, not elastically deformable and cannot be considered bi-dimensional. The body is not elastically deformable, and its vibration is strongly damped.

Acoustic radiation has been evaluated in two different ways in all musical instruments. The first step was to calculate the complex ratio (p/v) between sound pressure and vibration velocity of the principal vibration components of the instrument. This is the method used by Giordano.

In the second step, an amplitude of a space-averaged of cross spectrum ($p \cdot v$) between sound pressure and the vibrational speed of the membranes or chest was calculated at a fixed point 25 cm over the instruments. This is the parameter known as Intensity of Acoustic Radiation (IAR).

The IAR thus led to a parameter that was able to correctly associate the vibration of plate or membranes with the sound production. This relationship is remarkable for tympani, while the relationship is not so strong for *carabattola*. In the first situation, IAR and FRF are very correlated and only partly linked in the second situation. We have not considered the effect of the metal knocker, because our focus was in the vibroacoustic behavior of the soundboard.

Considering the modal patterns obtained from modal analysis and reported in Figure 5, it is noticeable that there are no clear modal patterns obtained from the analysis. This result is coherent with the sound behavior of the musical instrument, which is not characterized by a specific pitch but rather produces a strong peculiar noise which recalls a *grater*. Since there is not any specific pitch in the sound emission, there are no frequencies that emerged from the music spectrum. Moreover, the thickness of the soundboard, which is considerably higher than that of other wooden musical instruments or even plates, causes an early damping of the sound (or noise) emission that is compensated by the quick movement of the metallic handle, which hits the soundboard with a very high frequency.

In other words, the sound analysis carried out on the *carabattola* does not give evidence of any specific frequency but rather looks partially similar to a flatten spectrum without any decay, similarly to percussion instruments like a wooden scraper. Consequently, the modal patterns which resulted from the analysis resulted having an irregular shape, which is completely different from the membrane of the tympanum.

7. Conclusions

The *carabattola* is an almost unknown idiophone musical instrument that used to be played only during Holy Week in Mediterranean Europe. Few examples of this musical instrument are actually conserved in Italy, and no acoustic analyses are available for it. This paper attempted to describe its vibroacoustic behavior by means of some specific acoustic measurements: modal analysis and intensity of acoustic radiation. These methods have been already utilized in other musical instruments, as the kettledrum.

In the *carabattola*, comparing the FRF graph and *p/v*, it can be noted how frequencies of resonance often differ from those of acoustic emission. When IAR is applied, resonance frequencies completely match those of noise emissions, and the forms of the two graphs are comparable.

In the case of tympani, this was particularly evident: The IAR parameter was well related to the response function of frequencies, and therefore, it resulted preferable to *p/v*. IAR is a parameter between acoustic intensity and acoustic radiation, so it is suitable to measure the sound generating characteristics of musical instruments with vibrating soundboards (i.e., FRF). This parameter could also be used to describe and identify the directivity of musical instruments, essential both for architectural acoustics and for procedures of auralization.

In the case of the kettledrum, the links between FRF, IAR and modal analysis were very high, because of the clear subjective pitch of the M.I. In the case of the *carabattola*, where no pitch is detected, the correlations between FRF, IAR and modal patterns resulted very low. These results underline the link existing between IAR, modal patterns and pitch of the vibrating surface: Where no pitch is detected, low correlations occur between IAR and FRF, and modal analysis becomes less important.

The results obtained in the *carabattola* showed that modal patterns were not clearly identified as in the violin or kettledrum, and its sound emission was characterized by an almost uniformed frequency distribution, which typically characterizes idiophones.

Author Contributions: L.T. provided the funding, made the measurements and the results. M.M. contributed to the writing of the paper. V.V. finalized the paper including its formatting. All authors have read and agreed to the published version of the manuscript.

Funding: This research was funded by Italian Government in the framework of PRIN 2015. within the project "Research for SEAP: a platform for municipalities taking part in the Covenant of Mayors".

Acknowledgments: The Authors wish to thank Elisa Ferri for her precious help during the measurements and post processing.

Conflicts of Interest: The authors declare no conflict of interest.

References

1. Woodhouse, J. The acoustics of the violin: A review. *Rep. Prog. Phys.* **2014**, *77*, 115901. [CrossRef] [PubMed]
2. Askenfelt, A.; Jansson, E.V. From touch to string vibrations. I: Timing in the grand piano action. *J. Acoust. Soc. Am.* **1990**, *88*, 52–63. [CrossRef]
3. Askenfelt, A.; Jansson, E.V. From touch to string vibrations. II: The motion of the key and hammer. *J. Acoust. Soc. Am.* **1991**, *90*, 2383–2393. [CrossRef]
4. Farina, A.; Tronchin, L. On the "virtual" reconstruction of sound quality of trumpets. *Acust. United Acta Acust.* **2000**, *86*, 737–745.
5. Tronchin, L.; Manfren, M.; Vodola, V. Sound characterization through IAR measurement: A study of Persian Musical Instruments. *Appl. Sci.* **2020**, *10*, 633.
6. Tronchin, L. The Emulation of Nonlinear Time-Invariant Audio Systems with Memory by Means of Volterra Series. *J. Audio Eng. Soc.* **2012**, *60*, 984–996.
7. Farina, A.; Langhoff, A.; Tronchin, L. Acoustic characterisation of "virtual" musical instruments: Using MLS technique on ancient violins. *J. New Music Res.* **1998**, *27*, 359–379. [CrossRef]
8. Tronchin, L.; Coli, V.L. Further investigations in the emulation of nonlinear systems with Volterra series. *J. Audio Eng. Soc.* **2015**, *63*, 671–683. [CrossRef]
9. Carini, A.; Orcioni, S.; Terenzi, A.; Cecchi, S. Nonlinear system identification using Wiener basis functions and multiple-variance perfect sequences. *Signal Process.* **2019**, *160*, 137–149. [CrossRef]
10. Suzuki, H. Vibration and sound radiation of a piano soundboard. *J. Acoust. Soc. Am.* **1986**, *80*, 1573–1582. [CrossRef]
11. Giordano, N. Sound production by a vibrating piano soundboard: Experiment. *J. Acoust. Soc. Am.* **1998**, *104*, 1648–1653. [CrossRef]
12. Christian, R.S.; Davies, R.E.; Tubis, A.; Anderson, C.A.; Mills, R.I.; Rossing, T.D. Effects of air loading on tympani membrane vibrations. *J. Acoust. Soc. Am.* **1984**, *76*, 1336–1345. [CrossRef]

13. Quadrio, F.S. Della storia e della ragione d'ogni poesia volumi quattro di Francesco Saverio Quadrio. In *Milano, nelle stampe di Francesco Agnelli*; nelle stampe di Francesco Agnelli: Milano, Italy, 1741.
14. Spanu, G.N. SONOS-strumenti della musica popolare sarda Ilisso Editore. 1994. Available online: https://www.libreriauniversitaria.it/sonos-strumenti-musica-popolare-sarda/libro/9788885098305 (accessed on 15 January 2020). (In Italian)
15. Fahy, F.J. *Sound Intensity*; Elsevier: London, UK, 1989.
16. Wogram, K. Akustische Untersuchungen an Klavieren, in Der Piano -und Flÿgelbau Verlag Das Musikinstrument, Frankfurt; English version, Acoustical Research on Pianos: Vibrational Characteristics of the Soundboard. *Das Musikinstrument* **1984**, *24*, 380–404.
17. Tronchin, L. Modal analysis and intensity of acoustic radiation of the kettledrums. *J. Acoust. Soc. Am.* **2005**, *117*, 926–933. [CrossRef] [PubMed]
18. Piana, E.A.; Petrogalli, C.; Paderno, D.; Carlsson, U. Application of the wave propagation approach to sandwich structures: Vibro-acoustic properties of aluminum honeycomb materials. *Appl. Sci.* **2018**, *8*, 45. [CrossRef]

applied
sciences

Article

Sound Characterization through Intensity of Acoustic Radiation Measurement: A Study of Persian Musical Instruments

Lamberto Tronchin [1,*]**, Massimiliano Manfren** [2] **and Vincenzo Vodola** [1]

[1] Department of Architecture, University of Bologna, Via dell'Università 50, 47521 Cesena, Italy; vincenzo.vodola2@unibo.it

[2] Faculty of Engineering and Physical Sciences, University of Southampton, Highfields, Southampton SO17 1BJ, UK; M.Manfren@soton.ac.uk

* Correspondence: lamberto.tronchin@unibo.it

Received: 9 August 2019; Accepted: 3 January 2020; Published: 16 January 2020

Abstract: The physics of musical instruments often uses modal analysis as one of the most important methods for describing the behavior of sound chests and their acoustic radiation. However, some studies conducted in Europe (Wogram), Japan (Suzuki) and the United States (Giordano) underlined a very weak correlation between sound radiation and modal analysis. This mismatch required further research. The acoustic parameter intensity of acoustic radiation (IAR) is highly related to the mechanical vibration of the sound source. IAR was able to quantify the sound efficiency of musical instruments, and to relate sound radiation with modal analysis. This paper investigates IAR measured on three Persian stringed musical Instruments, namely the thar, si-thar and santoor. The analysis contributes to the knowledge of stringed musical instruments and sound propagation, since IAR is capable of linking the mechanical vibrations of soundboards with the sound generation of the musical instruments. The IAR results coming from the modal analysis carried out during the study of the instruments are reported herein.

Keywords: intensity of acoustic radiation; modal analysis; Persian musical instruments

---

1. Introduction

Several methods are available to describe and characterize the physics of musical instruments and their sound radiation. Ernst Chladni (1756–1827) first discovered that rigid and thin structures (e.g., plates), most of them symmetric and very clear, could depict special and particular mode shapes when they were bowed by a vibrating violin bow [1]. Chladni is therefore considered one of the pioneers (together with Robert Hooke) of modern experimental modal analysis. Starting from Chladni's studies, modal analysis became quite important for studies on musical instruments. Frederick A. Saunders, in his important works about stringed musical instruments' acoustics, described experimental modal analysis as one of the most important physical analyses of violins [2,3]. Carleen Hutchins, during her long activity as a scientist and violinmaker, considered modal analysis the most important and useful method for determining acoustic and sound quality in violin tops and backs [4,5]. However, no information about modal patterns and their importance (and rank) in the definition of sound characteristics was provided, although Hutchins stated that modes 2 and 5 were fundamental for the sound characterization of violin plates. These results were confirmed by several other authors, such as Erik V. Jannson [6,7]. Experimental modal analysis is based on the measurements of a frequency response function (FRF), which is obtained by exciting the sound board with a shaker or an impact hammer and capturing the vibration by means of accelerometers. Nevertheless, several authors such as Wogram [8], Suzuki [9] and Giordano [10] have not discovered a robust correlation between FRF and modal analysis. Moreover, they have questioned whether there was a rank among all the modes, and what the relationship is between modal analysis (i.e., mode shapes) and sound radiation. When Hutchins stated that modes 2 and 5 in violins were fundamental, she did not provide any scientific reason for not considering modes 1, 3 and 4; therefore, further research was required. Conversely, a different acoustic parameter that has recently been introduced called intensity of acoustic radiation (IAR) is well linked to the mechanical vibration of the sound source [11]. The IAR enables the sound characterization of musical instruments, including wind instruments [12].

Even though several measurements could be utilized for characterizing musical instruments, they do not enable the ranking of the different modal shapes or link modal shapes with sound radiation (i.e., sound production of musical instruments).

In this paper, the intensity of acoustic radiation (IAR) is experimentally evaluated for three Persian stringed musical instruments, namely the thar, si-thar and santoor. The IAR enables the determination of the most important frequencies responsible for sound propagation of the musical instruments. Furthermore, based on the results of the IAR, the modal patterns obtained from these frequencies are calculated. The Persian musical instruments were made in an original workshop in Tehran, Iran, then purchased and carried straight to Europe.

2. Materials and Methods

2.1. Acoustic Measurements on Musical Instruments

Apart from modal analysis, various other approaches have been introduced to study and evaluate the physical characteristics of musical instruments. Among these, we can distinguish physical and non-physical methods. Obviously, physical measurements can be of various kinds: mechanical, fluid-dynamic, chemical, electronic, and so on, and each one can contain subcategories (e.g., dendrochronology, mechanical resistance, C14, radiography, etc.). Acoustic measurements are therefore a small part of physical measurements, and are based essentially on two main phenomena: sound pressure (sound pressure, particle velocity, flow velocity) and mechanical vibration (acceleration, velocity, displacement). Consequently, the other main physical and acoustic measurements that can be identified are as follows:

- acoustic impedance (driving point mechanical impedance or admittance);
- directivity;
- sound intensity;
- damping (loss factor);
- acoustic radiation;
- finite element method (FEM);
- interferometry;
- frequency and impulse responses;

The last parameter allows the virtual emulation of the behavior of musical instruments by measuring impulse responses [13,14] as well as considering non-linear characteristics [15,16].

2.2. Sound Radiation from Musical Instruments

In order to determine the efficiency of the acoustic emissions of a surface that is vibrating, Fahy introduced the following relationship:

$$\sigma = \frac{W}{\rho_0 c S \langle \overline{v_n^2} \rangle} \tag{1}$$

where W represents the sound power radiated by a surface with area S, which could be obtained by integrating the far-field intensity over a hemispherical surface centered on the panel; $\langle \overline{v_n^2} \rangle$ represents the space-averaged value of the time-averaged normal distribution of velocity; and ρ_0 represents the air density and c the velocity of sound propagation (speed of sound).

The method introduced by Fahy could be also applied also to musical instruments, and it has been utilized by several researchers to the frame of musical instruments for calculating the possibilities of a soundboard to produce sound emission.

In 1984, Wogram introduced the parameter F/v, identifying F as the excitation vector and v as the corresponding velocity at the point of excitation (1984). Using the "surface-intensity" method (1986), Suzuki identified

$$I = Re[p(\alpha/j\omega*)/2] \tag{2}$$

where I represents the time average intensity, measured in the near acoustic field (about 30 cm from the radiating surface), perpendicular to the vibrating surface; and p and a represent the pressure and acceleration at the measuring point, respectively. Giordano used the parameter p/v, where p represents the sound pressure captured in the near field and v represents the velocity of the soundboard (1998). It is worthwhile considering that a similar method has been recently developed for the acoustic situation of sound insulation [17], far from musical instruments. Unfortunately, all these studies have one prevalent outcome: the frequencies of resonance have not coincided with those of acoustic emission, and they often have an adverse correlation.

2.3. Intensity of Acoustic Radiation—IAR

Starting from the studies described above, the intensity of acoustic radiation (IAR) parameter was defined as the space-averaged amplitude of the cross-spectrum between the sound pressure generated from the vibration of the surface and the velocity of the surface vibration. IAR is calculated by the equation

$$IAR(\omega) := \, < p(\omega) \cdot v(\omega) \tag{3}$$

where $p(\omega)$ represents the sound pressure measured in near field, and $v(\omega)$ represents the vibration velocity of the vibrating surface (i.e., the soundboard). The measurements of IAR require an omnidirectional microphone located in a fixed position in the near field, about 25–30 cm over the instruments, which should correspond to about one-fourth of the principal dimension of the instrument. Instead of a perfect anechoic room, the measurements should be carried out in a dry room that is almost anechoic (i.e., less than 0.2 s at mid frequencies), where average radiation induced by early modes (at low frequencies) is easily achieved via the reverberation time. The acoustics chamber has no effect on the readings at higher frequencies. In addition, the measurements are enhanced by the space-averaging of information carried out by shifting transductors through the instruments. The previous measurements, conducted by one of the authors on a kettledrum, have demonstrated the perfect matching between theoretical and experimental modal patterns. These results were also obtained during measurements on other musical instruments, as in Carabattola [18].

2.4. Persian Musical Instruments

Persian music utilizes a wide variety of musical instruments that differ from classical European ones in several aspects. Persian musical instruments could be regarded as similar to Indian instruments rather than European ones due to their features: many of them consist of a sound chest made with local wood, while the soundboard is often made from animal skin. In this paper, three distinct Persian stringed musical instruments, namely the thar, si-thar and santoor, were made in an original shop in Tehran, Iran, then purchased and carried straight to Europe. The scientific and historical literature on Persian musical instruments is very limited [19]. Moreover, only a few papers are available that discuss the acoustic behavior of them [20].

The thar (tār, tar, تار) is a Persian long-necked string instrument (Figure 1) that is commonly played in many countries on the Caucasus region. This musical instrument could be considered to be the most representative Persian musical instrument in view of its characteristic shape and sound. Often this musical instrument is also called a čāhārtār or čārtār, which means "four string" in Farsi. This is a common practice in Persian-speaking areas in which lutes are distinguished on the basis of the number of strings originally employed. Belonging to the lute family, the thar appeared in its present form in the middle of the eighteenth century and has since become one of the most important musical instruments in Iran and the Caucasus. The body is a double-bowl shape carved from mulberry wood, with a thin membrane of stretched lambskin covering the top. The long and narrow neck has a flat fingerboard with twenty-five to twenty-eight adjustable gut frets, and there are three double courses of strings. Its range is about two and one-half octaves (C3-A6), and it is played with a small brass plectrum.

Figure 1. The Persian thar analyzed in this research in its original case.

The si-thar (setâr, تارسه) is another important Persian musical instrument (Figure 2). It is a member of the lute family that is played with the index finger of the right hand. It has twenty-five to twenty-eight moveable frets that are usually made of gut. The si-thar covers more than two and a half scales (F4-G6). The ancestry of the si-thar can be traced to the ancient tanbur of Persia. As previously explained, following the rules of Persian speech, si-thar is literally translated as "three strings".

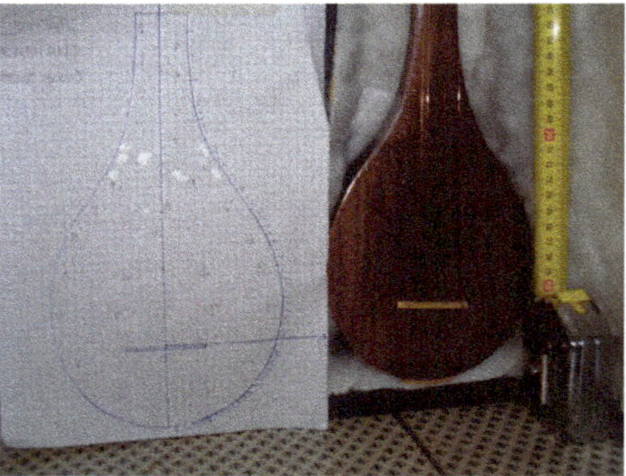

Figure 2. The Persian si-thar analyzed in this research in its original case.

The santoor (santur, رونتس) is a Persian three-octave (F3-G6) wooden-hammered dulcimer with seventy-two strings that are arranged on adjustable tuning pegs in eighteen quadruple sets, with nine (bronze) in the low register and nine (steel) in the middle register (Figure 3). The santoor can be made from various kinds of wood depending on the desired sound quality. However, there is no scientific information about which would be the preferable wood. The front and the back of the instrument are connected by soundposts whose positions play an important role in determining the sound quality of the instrument. Although the santoor is very old, it was neither depicted in miniatures, nor presented in any other medium until the 19th century.

Figure 3. The Persian santoor analyzed in this research.

To the authors knowledge, these three musical instruments have never been studied from an acoustic perspective, and this paper aims to evaluate their quality by means of acoustic measurements of IAR and modal analysis.

2.5. Acoustic Measurements

In order to characterize the sound and vibrational behaviors of these three Persian musical instruments, a measurement campaign was conducted. The measurements were made in an almost anechoic listening room (i.e., the Arlecchino room at University of Bologna, the same one in which the first experiments on kettledrums were performed in 2005) [11]. The room has a reverberation time lower than 0.2 s at 500 Hz, which helps to increase uniformity of sound radiation at low frequencies (i.e., at the first modes). At higher frequencies, the room acoustics do not influence the measurements. For all measurements, the equipment used was the following:

- Hammer Brüel & Kjær type 8203;
- Accelerometer Brüel & Kjær type 4374;
- two charge amplifiers Brüel & Kjær type 2635;
- PC equipped with 24 bit A/D converter soundcard and 8 channels with a 96 kHz sample rate;
- Soundfield MKV microphone;

Based on reciprocity theory, the accelerometer was located at the bridge in all of the instruments, in a few fixed positions, whilst the soundboards were excited by the hammer in dozens of positions.

3. Results

During the measurements, both acoustic pressure and velocity vibrations were collected and stored, as well as the excitation signal from the hammer. The modal analysis was undertaken for resonant frequencies up to 1 kHz. In addition, as described in Equation (3), the IAR was calculated in a second step for all three musical instruments.

3.1. Thar

Since the soundboard in the thar is produced of lamb skin, measuring vibrations became quite difficult. The thar has a wooden sound chest built of double-heart string instruments. The following impedance graph (Figure 4) displays an evaluation of the frequency response function along with the IAR.

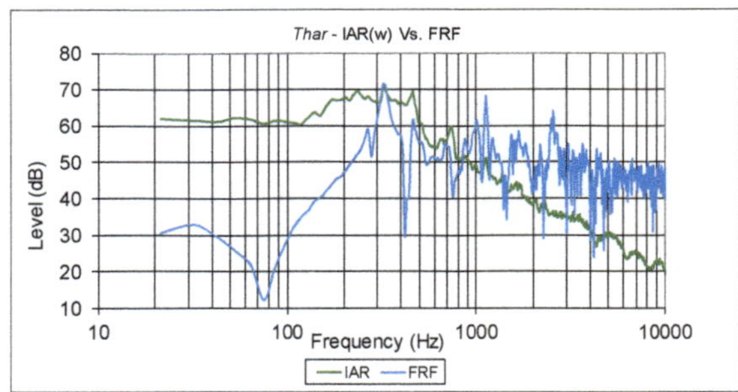

Figure 4. Intensity of acoustic radiation (IAR) compared with the frequency response function (FRF) for the Persian thar.

The following images (Figure 5) show the results of the modal analysis carried out during the analysis of acoustical characteristics of the thar.

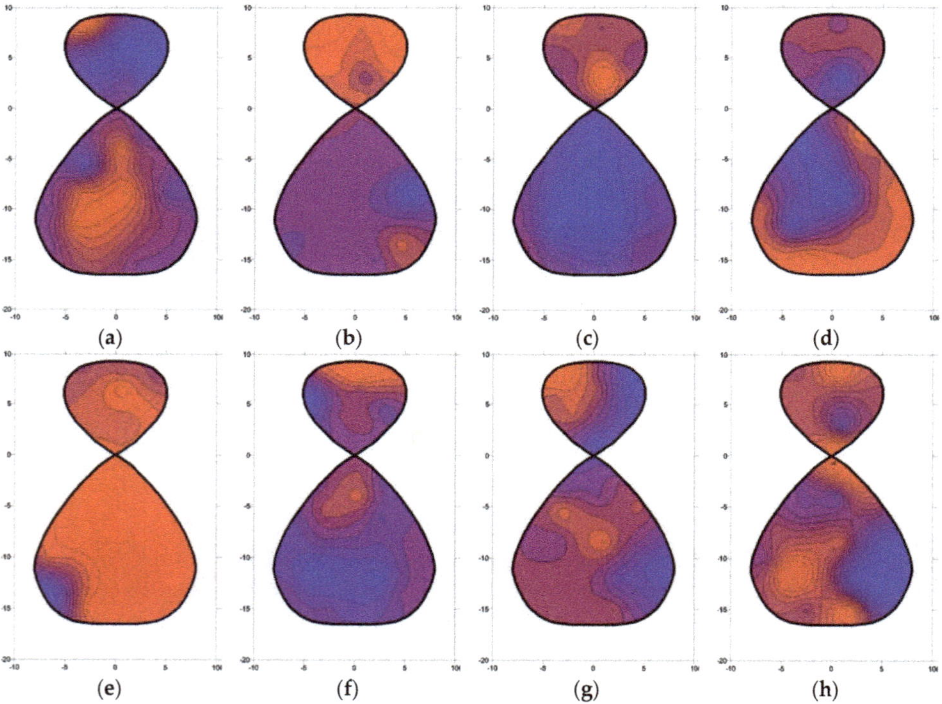

Figure 5. Results of modal analysis of the thar. Frequencies are (**a**) 32 Hz, (**b**) 269 Hz, (**c**) 322 Hz, (**d**) 462 Hz, (**e**) 592 Hz, (**f**) 689 Hz, (**g**) 882 Hz and (**h**) 1012 Hz.

3.2. The Si-thar

Due to its narrower size and lamb-skin soundboard, the Persian si-thar differs from the Indian one. As with the thar, the soundboard measurements required high precision and handling care. The following impedance graph (Figure 6) displays an evaluation of the frequency response function along with the IAR.

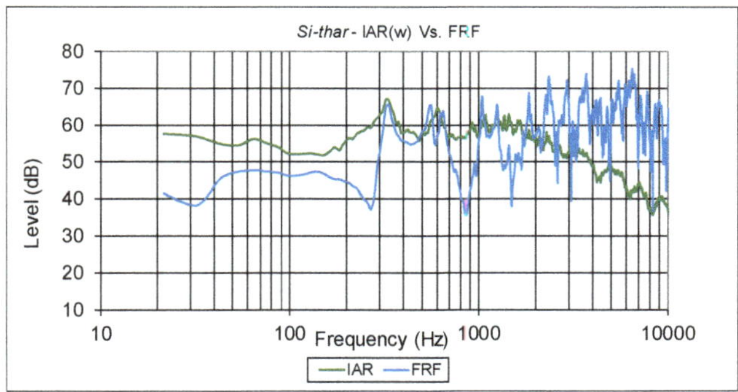

Figure 6. IAR compared with the FRF for the Persian si-thar.

Figure 7 shows the results relative to the modal analysis carried out during the analysis of acoustical characteristics of the si-thar.

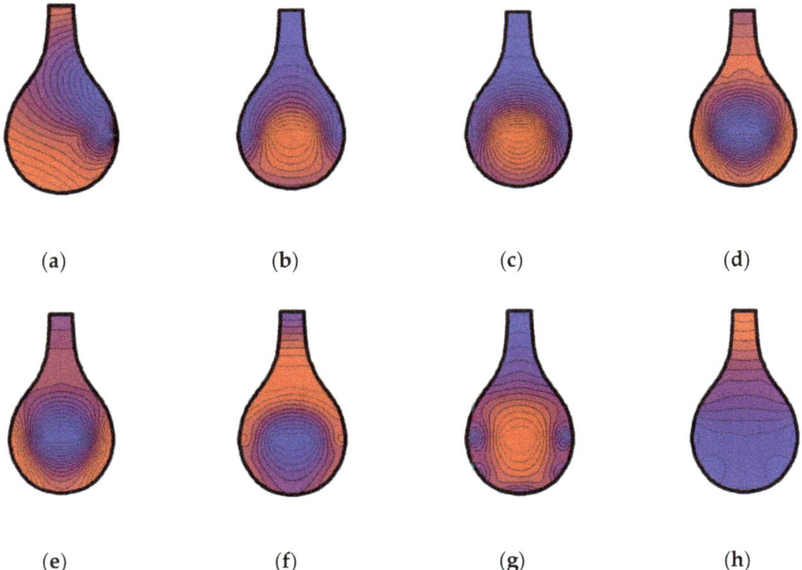

Figure 7. Results of modal analysis of the si-thar. Frequencies are (**a**) 53 Hz, (**b**) 139 Hz, (**c**) 172 Hz, (**d**) 226 Hz, (**e**) 269 Hz, (**f**) 322 Hz, (**g**) 549 Hz and (**h**) 645 Hz.

3.3. The Santoor

The santoor resembles a xylophone or clavichord, and it simpler to move than the other instruments. Although the soundboard is produced from wood and is resistant, it becomes almost impossible to measure soundboard vibrations in the sound chest due to the complexity of the musical instrument, which has a high number of strings and the feasibility of moving the bridge. Figure 8 show the impedance graph, which displays an evaluation of the frequency response function along with the IAR.

The modal analysis results obtained during the measurements are shown below (Figure 9).

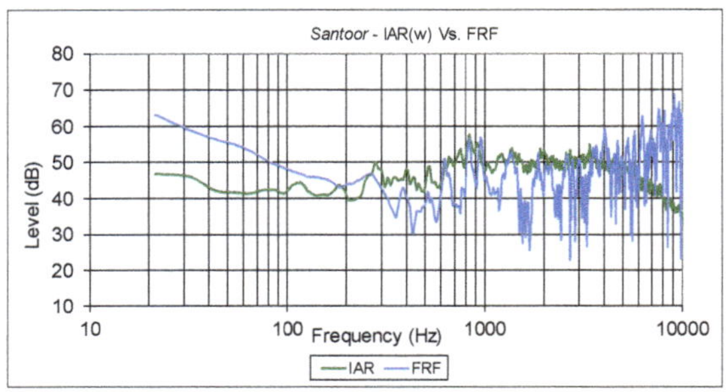

Figure 8. IAR compared with the FFR for the Persian santoor.

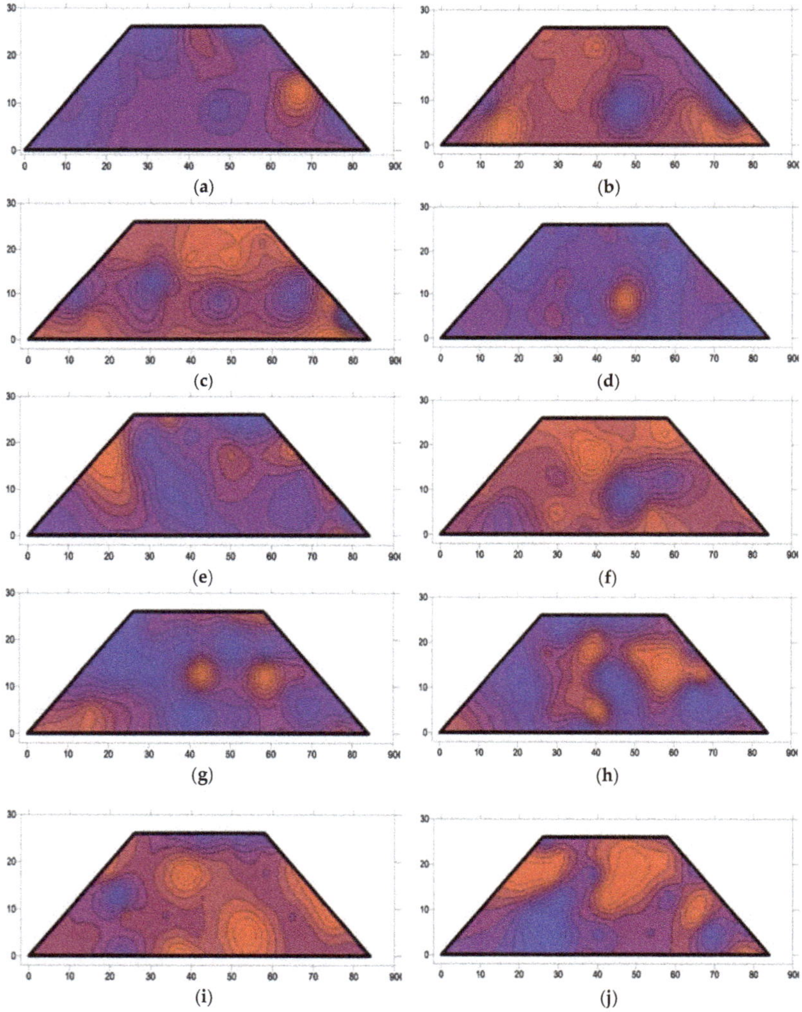

Figure 9. Results of modal analysis on the thar. Frequencies are (**a**) 269 Hz, (**b**) 387 Hz, (**c**) 516 Hz, (**d**) 538 Hz, (**e**) 624 Hz, (**f**) 785 Hz, (**g**) 839 Hz, (**h**) 958 Hz, (**i**) 1119 Hz and (**j**) 1378 Hz.

4. Discussion

Measurements of the IAR were reported in combination with FRF charts for all three musical instruments. In general, the low–mid IAR and FRF frequencies were near to each other in all cases, while correlations at the higher frequencies were lower. This result was already found for the kettledrum, and is a typical result of frequency analysis (with a narrow band of 10 Hz). Moreover, the thar and si-thar had a comparable frequency response. Both had a resonant frequency at 322 Hz, although they differed rather significantly at medium and high frequencies. Due to the large amount of cords and the more complicated form of the sound chest, the santoor generated a more balanced sound spectrum, which differentiated it from the other two string instruments.

By analyzing the results in more detail and comparing the IAR with modal analysis, we found the following:

In case of the thar, the most significative modal patterns that resulted were those at 322 and 462 Hz (i.e., a difference of 140 Hz). Moreover, if we consider the ratio between 322 and 140, it yields 2.3, whilst the ratio between 462 and 140 yields 3.3. This means that the ratio between the mode at 462 Hz and the mode at 322 Hz is 3:2 (i.e., a fifth), demonstrating that the dimensions of the two components of the soundboard allow a proper tuning of the frequency response of the musical instrument. This is not obvious, since the dimensions of the two components of the soundchest (i.e., the volume) are roughly in a 1:4 ratio.

In case of si-thar, analysis of the IAR showed that the two most relevant frequencies (i.e., the two main peaks) were 322 and 602 Hz, whereas for the FRF the most relevant frequencies among the 8 modal patterns reported here were 269, 322, 549 and 645 Hz. It could also be noted that 602 represents an important frequency for IAR, as it almost corresponds to the 592 Hz frequency at which the FRF has a minimum value and is between the two main values (549 and 645). Analyzing these frequencies, we see that, starting from 269 Hz while considering the value of 602 Hz for the IAR, they are in ratios of 6:5, 2:1, 18:8 and 12:5. This sequence represents almost a perfect minor natural (Zarlinian) scale, with 269–549 and 322–645 almost in the ratio of 2:1. This could be a reason for the typical melodies composed for this particular musical instrument.

The third musical instrument differed from the others for several reasons. For one thing, it is a percussion instrument, with a higher number of strings. Furthermore, the resonant cavity is much smaller and does not vibrate like the others, and the strings of the musical instrument are struck by little wooden hammers (mezrabs).

The analysis of IAR and modal patterns underlines the strong differences with the previous musical instruments. There was little evidence of specific resonant frequencies, even considering the IAR and FRF. The modal patterns were not very clear, unlike the previous cases. However, we could find a certain importance of the frequencies at 269 and 538 Hz (i.e., in the ratio 2:1), even if the corresponding modal pattern did not seem to underline a specific or noticeable shape. The other frequencies did not provide any specific evidence.

5. Conclusions

In this paper, the acoustic radiation and modal analysis of three Persian musical instruments was undertaken. The space-averaged amplitude of the cross spectrum $<p\ v>$ between sound pressure—measured at a fixed point of 25 cm from the instruments—and vibration velocity of the soundboard was calculated for several measuring points. This was the intensity of acoustic radiation parameter (IAR). Comparing the two graphs (i.e., the IAR and the FRF), some modal patterns for each musical instrument were obtained and reported. In the case of thar and si-thar, we found significant values of resonant frequencies that were in a harmonic ratio (3:2 for the thar; 2:1, 6:5, 18:8 and 12:5 for the si-thar). For each frequency, a clear modal pattern was obtained. In the case of the santoor, modal patterns were less evident than in the other musical instruments due to the complexity of the instrument, and we only found a significant value of frequencies in the ratio of 2:1. These results provide an initial description of the vibro-acoustic behavior of three important Persian musical instruments. Since the thar and si-thar have a similar vibroacoustic behavior, it is likely that other similar Persian (or Caucasian) stringed musical instruments might have the same sound characteristics. On the other hand, musical instruments like the santoor behave differently.

Author Contributions: L.T. provided the funding, made the measurements and the results. M.M. contributed to the writing of the paper. V.V. finalized the paper including its formatting. All authors have read and agreed to the published version of the manuscript.

Funding: This research was funded by Italian Government in the framework of PRIN 2015. Within the project "Research for SEAP: a platform for municipalities taking part in the Covenant of Mayors".

Acknowledgments: The authors wish to thank Fabrizio Dovesi for his precious help during the measurements and post processing, and Patrick James for his help in English proofing.

Conflicts of Interest: The authors declare no conflict of interest.

References

1. Chladni, E. Entdeckungen über die Theorie des Klanges. English version. In *Discoveries in the Theory of Sound*; Weidmanns Erben und Reich: Leipzig, Germany, 1787.
2. Saunders, F.A. The Mechanical Action of Violins. *J. Acoust. Soc. Am.* **1937**, *9*, 81–98. [CrossRef]
3. Saunders, F.A. Recent Work on Violins. *J. Acoust. Soc. Am.* **1953**, *25*, 491–498. [CrossRef]
4. Hutchins, C.M. A history of violin research. *J. Acoust. Soc. Am.* **1983**, *73*, 1421–1440. [CrossRef]
5. Hutchins, C.M. A 30-year experiment in the acoustical and musical development of violin-family instruments. *J. Acoust. Soc. Am.* **1992**, *92*, 639–650. [CrossRef]
6. Jansson, E.V. Experiments with violin plates and different boundary conditions. *J. Acoust. Soc. Am.* **1989**, *86*, 895–901. [CrossRef]
7. Jansson, E. Violin frequency response–bridge mobility and bridge feet distance. *Appl. Acoust.* **2004**, *65*, 1197–1205. [CrossRef]
8. Wogram, K. Akustische Untersuchungen an Klavieren, in Der Piano-und Flÿgelbau, Verlag Das Musikinstrument. English version. In *Acoustical Research on Pianos: Vibrational Characteristics of the Soundboard*; Das Musikinstrument: Frankfurt, Germany, 1984.
9. Suzuki, H. Vibration and sound radiation of a piano soundboard. *J. Acoust. Soc. Am.* **1986**, *80*, 1573–1582. [CrossRef]
10. Giordano, N. Sound production by a vibrating piano soundboard: Experiment. *J. Acoust. Soc. Am.* **1998**, *104*, 1648–1653. [CrossRef]
11. Tronchin, L. Modal analysis and intensity of acoustic radiation of the kettledrum. *J. Acoust. Soc. Am.* **2005**, *117*, 926–933. [CrossRef]
12. Bucur, V. *Handbook of Materials for Wind Musical Instruments*; Springer: Basel, Switzerland, 2019. [CrossRef]
13. Farina, A.; Langhoff, A.; Tronchin, L. Acoustic characterisation of virtual musical instruments: Using MLS technique on ancient violins. *J. New Music. Res.* **1998**, *27*, 359–379. [CrossRef]
14. Farina, A.; Tronchin, L. On the virtual reconstruction of sound quality of trumpets. *Acta Acust. United Acust.* **2000**, *86*, 737–745.
15. Tronchin, L. The Emulation of Nonlinear Time-Invariant Audio Systems with Memory by Means of Volterra Series. *J. Audio Eng. Soc.* **2012**, *60*, 984–996.
16. Tronchin, L.; Coli, V.L. Further Investigations in the Emulation of Nonlinear Systems with Volterra Series. *J. Audio Eng. Soc.* **2015**, *63*, 671–683. [CrossRef]
17. Piana, E.A.; Petrogalli, C.; Paderno, D.; Carlsson, U. Application of the Wave Propagation Approach to Sandwich Structures: Vibro-Acoustic Properties of Aluminum Honeycomb Materials. *Appl. Sci.* **2018**, *8*, 45. [CrossRef]
18. Tronchin, L.; Manfren, M.; Vodola, V. The *Carabattola*—vibroacoustical analysis and intensity of acoustic radiation (IAR). *Appl. Sci.* **2020**, *10*, 641. [CrossRef]

19.   Bouterse, C. Reconstructing the Medieval Arabic Lute: A Reconsideration of Farmer's Structure of the Arabic and Persian Lute. *Galpin Soc. J.* **1979**, *32*, 2–9. [CrossRef]
20.   Mansour, H.; Kasaiezadeh, A.; Arzanpour, S.; Behzad, M. Finite Element Modeling of Setar, a Stringed Musical Instrument. In *ASME 2009 International Mechanical Engineering Congress and Exposition*; IMECE: London, UK, 2009; pp. 591–597.

Article

A Study of the Dynamic Response of Carbon Fiber Reinforced Epoxy (CFRE) Prepregs for Musical Instrument Manufacturing

Manuel Ibáñez-Arnal *, Luis Doménech-Ballester and Fernando Sánchez-López

Department of Mathematics, Physics and Technology Sciences. University CEU Cardenal Herrera, CEU Universities, Avda. Seminario s/n, 46113 Moncada, Spain; luis.domenech@uchceu.es (L.D.-B.); fernando.sanchez@uchceu.es (F.S.-L.)
* Correspondence: manuel.ibanez@uchceu.es

Received: 23 September 2019; Accepted: 28 October 2019; Published: 30 October 2019

Abstract: Composite materials are presented in a wide variety of industrial sectors as an alternative to traditionally used materials. In recent years, a new sector has increasingly used these kinds of materials: the manufacture of musical instruments. Resonances of different elements that make up the geometries of musical instruments are commonly used with the aim of enhancing aspects of the timbre. These are sensitive to the mechanical characteristics of the material, so it is important to guarantee the properties of the composite. To do this, it is not uncommon to use pre-impregnated fibers (prepregs) which allow fine control of final volumetric fractions of the composite. Autoclaving is a high-quality process used to guarantee the desired mechanical properties in a composite, reducing porosity and avoiding delamination, but significantly raising production costs. On the contrary, manufacture without autoclaving increases competitiveness by eliminating the costs associated with autoclave production. In this paper, differences in dynamic behavior are evaluated under free conditions of different Carbon Fiber Reinforced Epoxy (CFRE) prepreg boards, processed by autoclave and out-of-autoclave. The results of the complex module are presented according to the frequency, quantifying the variations in the vibratory behavior of the material due to the change of processing.

Keywords: autoclave; out-of-autoclave; vacuum-bag-only; processing; CFRE; plates; modal; dynamic; musical instruments

---

1. Introduction

Interest in characterizing the elastic and damping properties of composites is increasing as more and more industrial sectors use these materials in very different applications. Among these, the most established sectors include the automotive, aerospace, and renewable energy industries [1–5]. However, in recent years, this has expanded to include other types of products, such as musical instruments.

In this expanding new sector, the modal and transient behavior of different structures has been studied. Commonly, in musical instruments resonances are usually sought to enhance the sound pressure levels obtained for different resonances, modifying the timbre and the qualitative aspects of the resulting sound. Research is particularly extensive in the chordophone family, particularly in the case of classical guitars [6–12], electric guitars [13,14], or the string family of instruments [15–19]. We also find studies linked to percussive chordophones such as the piano [20–22] and, though scarcer, those dedicated to membranophones [23–25]. Figure 1 shows some examples.

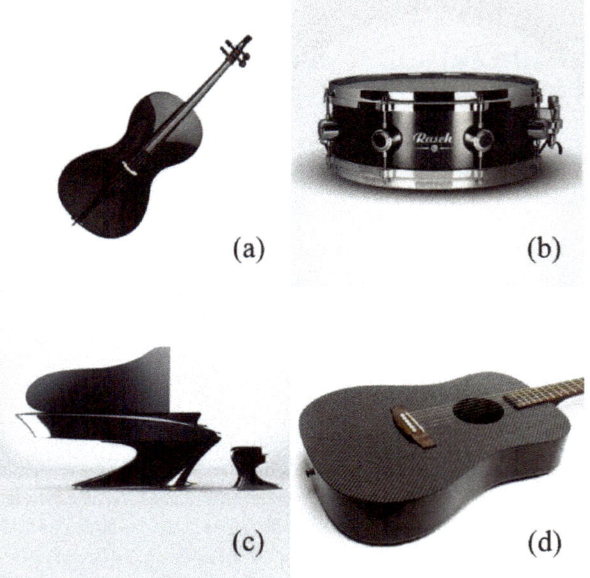

Figure 1. Some musical instruments made of carbon fiber reinforced with epoxy (CFRE): (**a**) Luis and Clark violoncello [26], (**b**) Rasch snare drum [27], (**c**) Boganyi piano [28], and (**d**) Klos carbon fiber guitar [29].

Composite materials use polymer matrices in their manufacture, so they have viscoelastic behavior. Viscoelasticity is the property of materials that exhibit both viscous and elastic characteristics when undergoing deformation. For the study of this type of material, we express the module in its complex form, as shown by Equation (1) [30]

$$E^* = E' + jE''$$ (1)

where the real part (E') is the storage modulus, and the imaginary part (E'') is the loss modulus.

The study of the storage module (E') is a determining factor when characterizing the resonance frequencies of a solid. E' represents the elastic behavior of the material, which is directly linked to the resonance frequencies of any type of solid. Also, for typical resonant elements of musical instruments, such as membranes, plates or shells, we can express their resonances as [31]

$$f = \alpha \sqrt{\frac{E'}{\rho}}$$ (2)

where the square root of the ratio between E' and ρ is equivalent to the longitudinal speed of sound in the solid, and α is a parameter defined by the modal and geometric characteristics of the solid. Variations of E' modify the value of each one of the natural frequencies of the system.

In turn, the modulus of losses E'', is related to the total damping of the system by means of Equation (5) and defines the amount of energy dissipated in each oscillation. The non-elastic or viscous behavior is related to the amount of sound pressure that a resonance will be able to accumulate and the modal decay time. The decay time due to internal damping processes is defined by [32]:

$$\tau = \frac{1}{\pi f} \frac{E'}{E''}$$ (3)

Because of their configurability, many manufacturers are investigating the use of composite materials for the manufacture of these instruments. The most common are those using Carbon Fiber Reinforced Epoxy (CFRE) [11,33,34], but others include those involving bio-composites [35,36]. These studies relate the mechanical characteristics of the composite to the sound behavior of different musical instruments, but do not contemplate the direct influence of the processing on the elastic and non-elastic behavior of the vibration.

Since the vibratory behavior and mechanical properties of the material are closely linked [32,37–39], in many cases the processing of the composite offers greater guarantees, such as the laminate of prepregs in an autoclave [40]. But due to the high associated cost, there are numerous studies that propose vacuum bag-only (VBO) processing as an alternative. Previous work has focused on investigating the effects that this can have on porosity, bubbles during curing, and different strategies for mitigating the defects associated with VBO [41–44], while trying to retain the mechanical properties that the autoclave offers.

It is known that defectology in processing affects the properties of the composite [45]. There are studies linking the appearance of voids during the processing of the composite to the mechanical characteristics of the composite [45,46], and others relating them to material failure [47]. Figure 2 shows the effects of defectology associated with VBO processing [48].

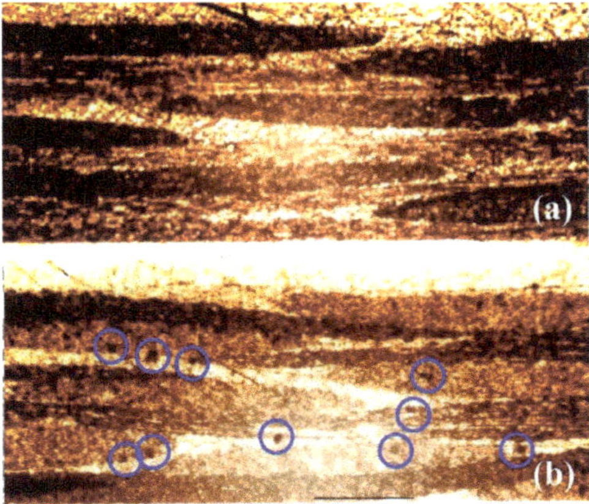

Figure 2. Micrographs of laminates: (**a**) autoclave processing, and (**b**) vacuum bag-only (VBO) processing. Blue circles show voids in the VBO laminate.

Studies have also investigated dynamic behavior in composites based on various variables, such as the addition of viscoelastic layers [49] or optimization of damping by characterizing laminate [50]. Since the vibrations of the material are of reduced amplitudes (<0.1 mm) in the musical instrument manufacturing industry, we can focus the study of the material only on the type of resonances emitted, without addressing aspects such as the failure of the material, either from fatigue or another mechanism.

Accordingly, it is essential to study the direct influence of processing on the modal behavior of the CFRE in so far as different manufacturers can establish criteria for choosing the type of processing. Therefore, this paper aims to evaluate the magnitude of the direct influence of processing by prepreg CFRE, autoclave or vacuum bag-only, on the vibratory characteristics of the composite, considering its sound speed, obtained by its modal behavior, as well as in the decay time result of the damping of the laminate.

2. Materials and Methods

For this study two laminates were generated, one by autoclave (A) and the other by vacuum bag-only (VBO) as shown in Figure 3.

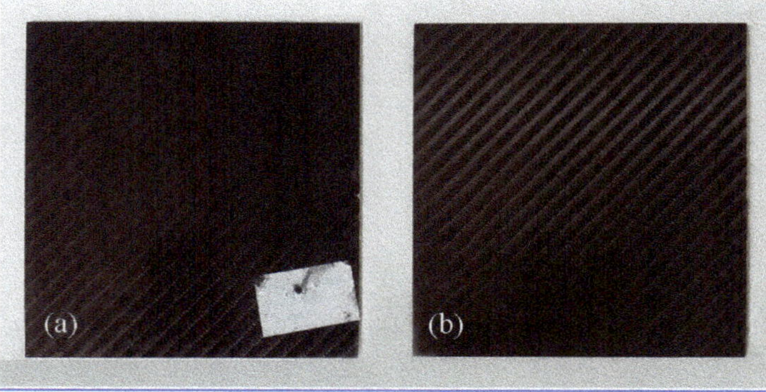

Figure 3. (**a**) Autoclave and (**b**) VBO CFRE plates.

Since each solid presents resonances that can be expressed as a direct proportion to the speed of sound of the material that constitutes it, we can extrapolate from the study of the speed of sound to a simple geometry that allows analysis of the material with precision. The geometry used is the rectangular plate, since it allows for obtaining a great number of bending vibration modes.

One way to express the resonances of a rectangular plate is [32]

$$f_n = \frac{0.113h}{L^2} \sqrt{\frac{E}{\rho}} \, [2n + 1] \qquad (4)$$

where h is the thickness, L is the length of the plate, E is the Young modulus, ρ is the density, and n is the number of nodes.

To be able to cover a range equivalent to three octaves above and below the A3 reference (440 Hz), the frequency range of the study was focused such that $0 < f < 2000$ Hz. For this, the plate dimensions must be adjusted using Equation (4), which defines the resonances for plates.

The multipurpose CFRE system considered for this investigation was GG280T (Tenax HTA-3k)-DT806R-42 Fabric Laminate manufactured by an external company, (Magma Composites S.L, Alcañiz, Aragón, Spain) using prepreg plies supplied by Delta Preg Composites. DT806 is an epoxy resin with a glass transition temperature (Tg) of approximately 135 °C. GG280T is a 4/4 twill carbon fabric, 3K high strength (HS) carbon fiber reinforcement having a density of 198 g/cm^2. Plates with dimensions of 220 × 220 mm were manufactured by the vacuum bag-only and the autoclave method, layering three collinear plies to an approximate total thickness of 1.03 mm. Plate A was cured in an autoclave at 120 °C for 1 hour, without post-cure. The pressure was 4 atm for the autoclave plate (A), and atmospheric pressure was used for the VBO plate. The manufactured plates were afterwards cut down to 30 × 200 mm plates, as shown in Figure 4.

A total of five specimens (dimensions: 200 mm × 30 mm × 1.035 mm) were manufactured using three collinear layers, considering that the fiber directions were parallel to the sides of the rectangles that make up the specimen, as shown in the Figure 5.

Properties provided by the manufacturer are shown in Table 1.

In order to characterize the elastic and non-elastic properties of the specimens, a dynamic test was carried out using the vibration procedure under free conditions stipulated in ISO 6721-3 [45], as illustrated in Figure 6.

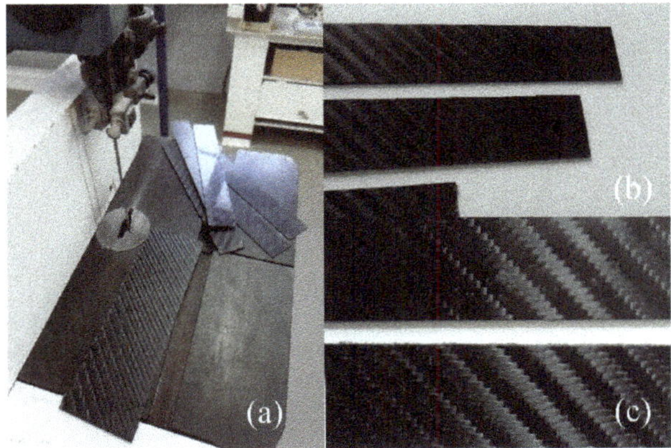

Figure 4. Manufacturing process of CFRE specimens: (**a**) cutting of 30 × 200 mm specimens; (**b**) finished specimens; and (**c**) detailed visual comparison of the surface of autoclave and VBO top surfaces.

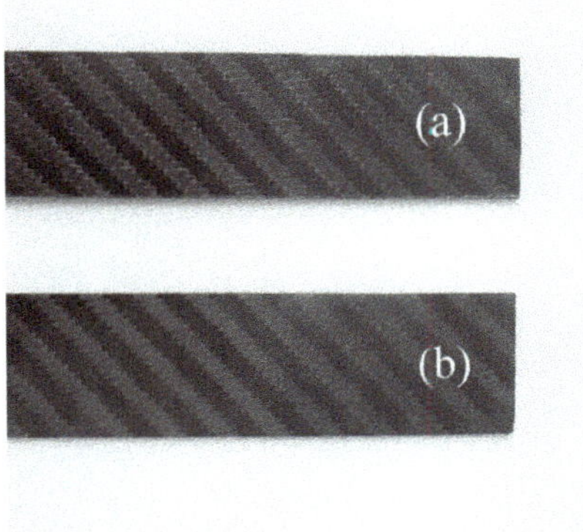

Figure 5. Specimens (**a**) VBO and (**b**) autoclave.

Table 1. Static mechanical properties of carbon fiber reinforced epoxy prepreg.

Density (kg/m³)	Young Modulus (GPa)			Poisson Ratio			Shear Modulus (GPa)		
	x	y	z	Xy	yz	xz	xy	yz	xz
1420	35.9	35.9	6.9	0.04	0.34	0.34	11.5	2.7	2.7

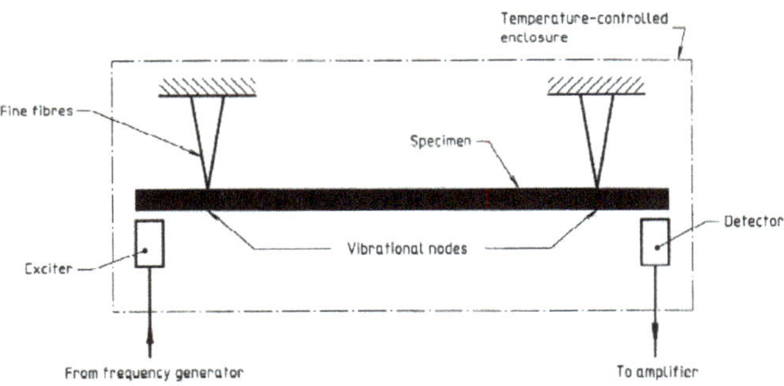

Figure 6. Experimental procedure for identifying resonance frequencies by external excitation [51].

In order to obtain experimental resonances for each plate, a wave generator was used to generate a frequency sweep for $0 < f < 2000$ Hz. This sine wave was transferred to an amplifier, which sent the signal to a coil exciting one end of the plate. On the opposite side of the plate, a flat-spectrum microphone was used to capture the sound pressure levels that the plate emits at different frequencies, identifying the resonances of each of the plates.

Once all the resonance frequencies were obtained, the signal decrement was captured. To do this, the plate was subjected to resonance, and the external excitation was cut at the moment of greatest amplitude to let the plate oscillate freely until the oscillation stops. With all captured data, the following calculations were performed.

Vibration modes were obtained as Figure 7 shows, thus obtaining the values of the storage modulus for each of the resonances $⟦E'⟧\_f$ with Equation (5). Resonance frequencies for a total of five vibration modes (at pure bending) shown in Figure 8.

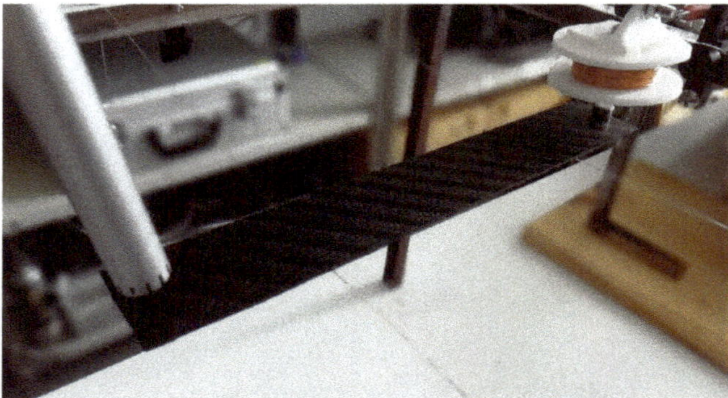

Figure 7. Experimental test for the modal analysis of prepreg CFRE plates.

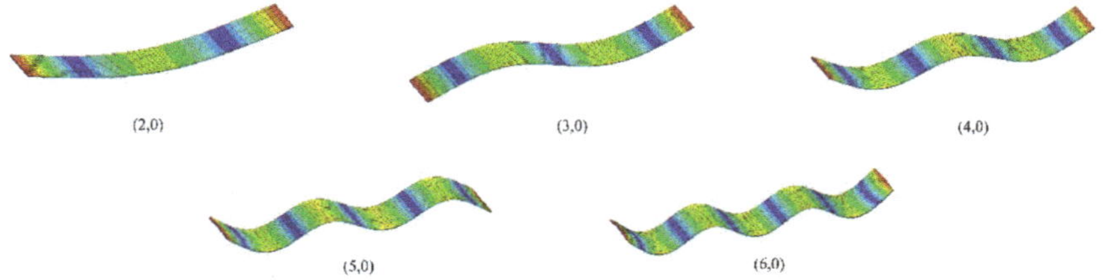

Figure 8. Modes obtained through experimentation, following the nomenclature (m,n) [32].

The storage modulus is defined as

$$E'_f = \left[\frac{4\pi(3\rho)^{1/2}l^2}{h} \right]^2 \left(\frac{f_{ri}}{k_i^2} \right)^2 \tag{5}$$

where ρ is the density, l is the length of the specimen, h is the thickness of the plate, f_{ri} corresponds to the resonance frequency of the i-th mode, and k_i^2 is a numerical factor dependent on the i-th vibrational mode and the contour conditions of the plate [45].

E'_f allows the calculation of the speed of sound in the composite, by its relationship to density, using the well-known Equation (6):

$$c = \sqrt{\frac{E'}{\rho}} \tag{6}$$

In the second part of the study, the composite loss modulus E''_f was obtained for each of the vibrational modes.

To do this, the signal decrement of the oscillations in free conditions was analyzed for each of the resonances obtained.

The logarithmic decrement is defined as

$$\delta = ln\left(\frac{A_1}{A_n}\right)/N \tag{7}$$

where A_1 is the amplitude of the initial oscillation, and is the amplitude of the n-th oscillation after the start of the decay.

The loss factor can be expressed as

$$E''_f = E'_f \tan\phi_f \tag{8}$$

where ϕ_f is the phase or loss angle.

There is a direct relationship between the phase and the loss factor [52]; for $\phi \ll 1$ we can obtain using the following equation:

$$\tan\phi_f = \frac{\delta_f}{\pi} \tag{9}$$

3. Results and Discussion

3.1. Dynamic Characterization

The results obtained for each of the processed specimens were then analyzed using the procedure described above, as shown in Tables 2 and 3.

Table 2. Average values obtained for CFRE prepreg plates processed by autoclave.

$\rho\,(Kg/m^3)$	Mode	f (Hz)	E' (Pa)	E' (Pa)	η	ϕ (Rad)	δ	c (m/s)
1.39×10^3	2.0	1.17×10^2	2.73×10^{10}	6.12×10^7	2.24×10^{-3}	2.24×10^{-3}	7.05×10^{-3}	4.44×10^3
	3.0	3.52×10^2	3.19×10^{10}	7.85×10^7	2.46×10^{-3}	2.46×10^{-3}	7.74×10^{-3}	4.80×10^3
	4.0	7.09×10^2	3.37×10^{10}	9.18×10^7	2.73×10^{-3}	2.73×10^{-3}	8.57×10^{-3}	4.93×10^3
	5.0	1.17×10^3	3.36×10^{10}	1.02×10^8	3.04×10^{-3}	3.04×10^{-3}	9.55×10^{-3}	4.92×10^3
	6.0	1.69×10^2	3.12×10^{10}	1.28×10^8	4.10×10^{-3}	4.10×10^{-3}	1.29×10^{-2}	4.74×10^3

Table 3. Average values obtained for CFRE prepreg plates processed by vacuum bag-only.

$\rho\,(Kg/m^3)$	Mode	f (Hz)	E' (Pa)	E'' (Pa)	η	ϕ (Rad)	δ	c (m/s)
1.16×10^3	2.0	1.15×10^2	2.00×10^{10}	5.18×10^7	2.59×10^{-3}	2.59×10^{-3}	8.14×10^{-3}	4.15×10^3
	3.0	3.38×10^2	2.25×10^{10}	5.60×10^7	2.49×10^{-3}	2.49×10^{-3}	7.81×10^{-3}	4.40×10^3
	4.0	6.78×10^2	2.37×10^{10}	3.86×10^8	1.63×10^{-2}	1.63×10^{-2}	5.12×10^{-2}	4.51×10^3
	5.0	1.11×10^3	2.31×10^{10}	3.93×10^8	6.60×10^{-2}	6.59×10^{-2}	2.07×10^{-1}	4.46×10^3
	6.0	1716.18	2.48×10^{10}	3.91×10^8	1.58×10^{-2}	1.58×10^{-2}	4.95×10^{-2}	4.62×10^3

We can see some notable differences between the two types of processing. As Figure 9 shows, the processing change caused a 27% decrease in the real part of the E'_f module, which was practically constant for all frequency values. We also observe in the results how the mean values of equivalent density of the plate decrease markedly due to the effects of porosity; the test processed using VBO had a density 16.2% lower than that processed by A.

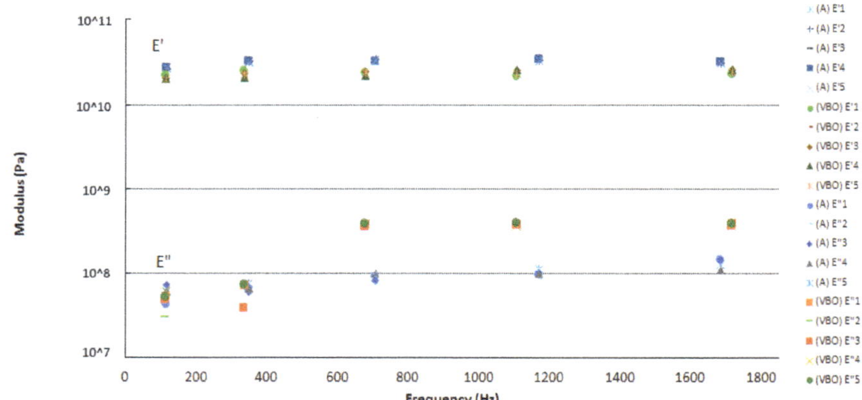

Figure 9. Loss and storage modulus results for CFRE prepregs: A (autoclave) and VBO (vacuum bag-only).

Although both values decrease markedly, we see how the mean frequency values were less than each for both processes, with variations of between 2%–6%.

This can be explained by the resulting sound speed of composite c defined by Equation (6) and represented in Figure 10. It is defined as the square root of the ratio between the stiffness and the density of the medium. Because both variables are diminished by porosity, the ratio between them is maintained by mitigating the effect of processing losses, maintaining the wave rate and therefore the values for the resonance frequencies that are proportional to their sound speed [30].

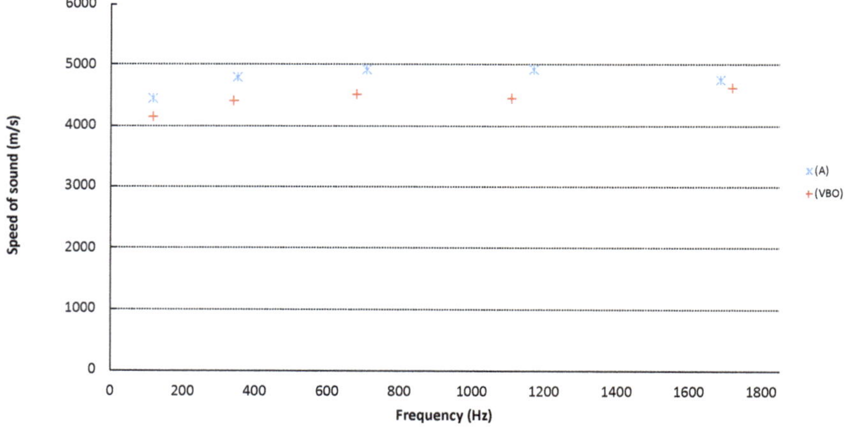

Figure 10. Average sound speed (c) at different frequencies, for CFRE prepregs, processed by A (autoclave) and VBO (vacuum bag-only).

The biggest changes are seen in the non-elastic behavior of the composite. As Figure 9 shows, VBO processing generated a large increase in the loss modulus E''_f especially in the range $f > 400$ Hz; for example, 4.2 times higher for vibration mode (4.0) located around 700 Hz.

The increase of the loss modulus against the storage module generated an increase in damping, or loss factor η, of the composite processed by VBO. As shown, this increase was maximized for 1100 Hz, completely changing the behavior of the material that, in the case of the plate processed by autoclave, exhibited a behavior almost proportional to the frequency.

3.2. Effects on The Vibrational Behavior.

The effect of processing was clearly identifiable in transient analysis Figure 11. As Figure 12 shows, for vibration mode 2.0 with a frequency of $f \cong 115$ Hz the decay times are similar. However, for vibration mode 5.0 with a frequency $f \cong 1150$ Hz, damping values were much higher for the plate processed out-of-autoclave, so the duration of the vibration is much shorter.

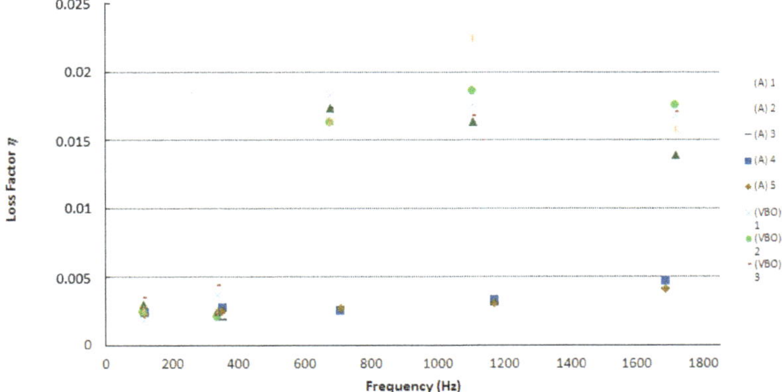

Figure 11. Comparison of the loss factor for CFRE prepregs processed by autoclave and vacuum bag-only.

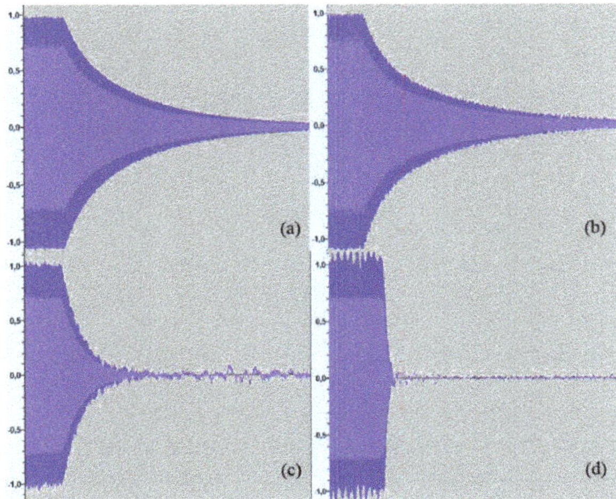

Figure 12. Transient analysis of CFRE plates: (**a**) Autoclave 2.0 vibrational mode; (**b**) vacuum bag-only 2.0 vibrational mode; (**c**) autoclave 5.0 vibrational mode; and (**d**) vacuum bag-only 5.0 vibrational mode.

The musical instrument manufacturing industry is commonly based on the use of resonances of secondary elements to boost the sound pressure levels of elements such as strings and membranes, but also to emit sound when impacted, as is the case with idiophones.

In both cases, the frequency values for each of the resonances are important, in addition to the damping values, since they define the amount of amplitude that each resonance can accumulate. Higher damping values result in a dissipation of the vibration energy.

As shown in Figure 13, the effect of the different processes of the CFRE is noticeable in the vibratory response of the composite.

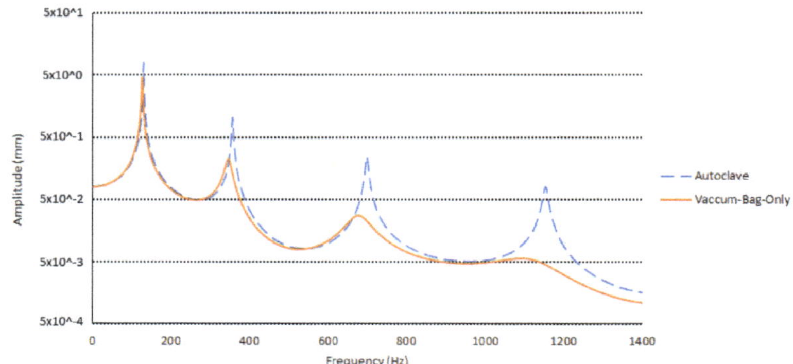

Figure 13. Maximum amplitudes recorded on the different CFRE plates by frequency sweep excitation for $0 < f < 1400$ Hz.

The values of the resonance frequencies were stable due to the similarity of the sound velocities in both processes. The biggest differences were due to the non-elastic behavior of the plates. For low frequency values, where the mechanical differences between the plates are minimal, the resonances were practically equal in amplitude and frequency.

As the frequency increases, the effects of the damping on the VBO plate become more important. This effect is critical near 1000 Hz, where some resonances disappear and their capacity to accumulate amplitude decreases.

4. Conclusions

This paper performed a dynamic test to analyze the speed of sound of CFRE prepreg plates processed by two different processes, autoclave and vacuum bag-only. In line with previous studies, we find that processing is an important factor to consider when dynamically characterizing a CFRE prepreg and is undoubtedly a new variable to be taken into account in material design applied to the musical instrument manufacturing industry.

However, although mechanical and physical differences were quantified in the resulting material, due to the stability of the sound speed in the composite, the vacuum bag-only method in CFRE prepregs is shown to be a feasible process for all those applications which require characterizing the elastic behavior and resonance frequencies of a product. Opting for an out-of-autoclave process is guaranteed to be sufficient for the manufacture of musical instruments, if we attend to the values of the resonance frequencies.

The biggest differences were generated at the non-elastic level; there were very significant increases in composite damping, which is something to consider in the design and production process. High damping values mean oscillation energy dissipation, so resonances accumulate less sound pressure. This factor may represent an added difficulty in the musical instrument manufacturing industry, where it is primarily of interest to the maximum accumulation of vibration energy to boost sound pressure levels, and generally materials with low attenuation are suited to this purpose.

In both cases, the knowledge of the influence of processing on the CFRE prepreg allows a more precise characterization of the material. While both processes can ensure stable resonances, due to their similarly elastic behavior, autoclave processing offers composites with lower damping values, improving maximum sound pressure levels. Thus, depending on the frequency range, the use of autoclave processing remains interesting.

Author Contributions: Conceptualization and writing—original draft preparation, M.I.-A.; methodology, M.I.-A. and validation, M.I.-A., F.S.-L. and L.D.-B.; investigation, M.I.-A.; writing—original draft preparation, M.I.-A.; writing—review and editing, F.S.-L., L.D.-B.; supervision, F.S.-L.

Funding: This research received no external funding.

Conflicts of Interest: The authors declare no conflict of interest.

References

1. Lee, H.G.; Lee, J. Damping mechanism model for fatigue testing of a full-scale composite wind turbine blade, Part 2: Application of fairing. *Compos. Struct.* **2018**, *202*, 1121–1125. [CrossRef]
2. Lee, H.G.; Lee, J. Damping mechanism model for fatigue testing of a full-scale composite wind turbine blade, Part 1: Modeling. *Compos. Struct.* **2018**, *202*, 1216–1228. [CrossRef]
3. Henning, F.; Kärger, L.; Dörr, D.; Schirmaier, F.J.; Seuffert, J.; Bernath, A. Fast processing and continuous simulation of automotive structural composite components. *Compos. Sci. Technol.* **2019**, *171*, 261–279. [CrossRef]
4. Chen, Y.; Wu, H.; Zhai, J.; Chen, H.; Zhu, Q.; Han, Q. Vibration reduction of the blisk by damping hard coating and its intentional mistuning design. *Aerosp. Sci. Technol.* **2019**, *84*, 1049–1058. [CrossRef]
5. Tekin, E.; Kapan, Ö. Composite Manufacturing Data Management in Aerospace Industry. *Procedia CIRP* **2016**, *41*, 1039–1042. [CrossRef]
6. Sumi, T.; Ono, T. Classical guitar top board design by finite element method modal analysis based on acoustic measurements of guitars of different quality. *Acoust. Sci. Technol.* **2008**, *29*, 381–383. [CrossRef]
7. Torres, J.A.; Boullosa, R.R. Influence of the bridge on the vibrations of the top plate of a classical guitar. *Appl. Acoust.* **2009**, *70*, 1371–1377. [CrossRef]
8. Bilbao, S.; Torin, A.; Chatziioannou, V. Numerical modeling of collisions in musical instruments. *Acta Acust. United Acust.* **2015**, *101*, 155–173. [CrossRef]
9. Caldersmith, G. Designing a guitar family. *Appl. Acoust.* **1995**, *46*, 3–17. [CrossRef]
10. Elejabarrieta, M.J.; Santamaría, C.; Ezcurra, A. Air cavity modes in the resonance box of the guitar: The effect of the sound hole. *J. Sound Vib.* **2002**, *252*, 584–590. [CrossRef]
11. Ono, T.; Isomura, D. Acoustic characteristics of carbon fiber-reinforced synthetic wood for musical instrument soundboards. *Acoust. Sci. Technol.* **2004**, *25*, 475–477. [CrossRef]
12. Torres, J.A.; Torres-Torres, D. Cambios en la propagación de ondas en una tapa de guitarra debidos al abanico y el puente. *Rev. Int. Métodos Numér. Cálc. Diseño Ing.* **2015**, *31*, 228–234. [CrossRef]
13. Nishimura, K.; Nishimura, K. A study on timbre and sound quality of an electric guitar by selection of material around pickup. In Proceedings of the ICSV 23rd International Congress on Sound and Vibration: From Ancient to Modern Acoustics, Athens, Greece, 10–14 July 2016.
14. Paté, A.; Le Carrou, J.L.; Fabre, B. Modal parameter variability in industrial electric guitar making: Manufacturing process, wood variability, and lutherie decisions. *Appl. Acoust.* **2015**, *96*, 118–131. [CrossRef]
15. Knott, G.A.; Shin, Y.S.; Chargin, M. A modal analysis of the violin. *Finite Elem. Anal. Des.* **1989**, *5*, 269–279. [CrossRef]
16. Yu, Y.; Jang, I.G.; Kim, I.K.; Kwak, B.M. Nodal line optimization and its application to violin top plate design. *J. Sound Vib.* **2010**, *329*, 4785–4796. [CrossRef]
17. Inácio, O.; Antunes, J.; Wright, M.C.M.M. Computational modelling of string–body interaction for the violin family and simulation of wolf notes. *J. Sound Vib.* **2008**, *310*, 260–286. [CrossRef]
18. Corradi, R.; Liberatore, A.; Miccoli, S. Experimental modal analysis and finite element modelling of a contemporary violin. In Proceedings of the ICSV 23rd International Congress on Sound and Vibration: From Ancient to Modern Acoustics, Athens, Greece, 10–14 July 2016.
19. Bissinger, G. Modal analysis of a violin octet. *J. Acoust. Soc. Am.* **2003**, *113*, 2105–2113. [CrossRef]
20. Berthaut, J.; Ichchou, M.N.; Jézéquel, L. Piano soundboard: Structural behavior, numerical and experimental study in the modal range. *Appl. Acoust.* **2003**, *64*, 1113–1136. [CrossRef]
21. Boutillon, X.; Ege, K. Vibroacoustics of the piano soundboard: Reduced models, mobility synthesis, and acoustical radiation regime. *J. Sound Vib.* **2013**, *332*, 4261–4279. [CrossRef]
22. Ege, K.; Boutillon, X.; Rébillat, M. Vibroacoustics of the piano soundboard: (Non)linearity and modal properties in the low- and mid-frequency ranges. *J. Sound Vib.* **2013**, *332*, 1288–1305. [CrossRef]
23. Ibáñez, M.; Sánchez, F. Material characterization and vibro-acoustic analysis of a Preinpregnated Carbon Fiber reinforced epoxy Drum Shell. In Proceedings of the 20th International Conference on Composite Materials, Copenhagen, Denmark, 19–24 July 2015.

24. Ibañez, M.; Muñoz, E.; Domenech, L.; Cortés, E.; Sánchez, F.; García, J.A. On the influence of mechanical and processing characterization on the vibro-acoustic response of lcm and preimpregnated composite laminates. In Proceedings of the 13th International Conference on Flow Processes in Composite Materials, Kioto, Japan, 6–9 July 2016.
25. Sathej, G.; Adhikari, R. The eigenspectra of Indian musical drums. *J. Acoust. Soc. Am.* **2009**, *125*, 831–838. [CrossRef] [PubMed]
26. Luis and Clark Carbon Fiber Instruments. Available online: https://luisandclark.com (accessed on 12 October 2019).
27. Rasch Drums. Available online: www.raschdrums.com (accessed on 10 October 2019).
28. Boganyi Piano. Available online: http://www.boganyi-piano.com (accessed on 12 October 2019).
29. Klos Carbon Fiber Guitars. Available online: https://klosguitars.com (accessed on 12 October 2019).
30. Nilsson, A.; Liu, B. *Vibro-Acoustics*; Springer: Berlin/Heidelberg, Germany, 2012; Volume 1, ISBN 9783662478066.
31. Chaigne, A.; Campbell, M. *Acoustics of Musical Instruments*; Springer: Berlin/Heidelberg, Germany, 2016; ISBN 9781493936779.
32. Fletcher, N.H.; Rossing, T.D. *The Physics of Musical Instruments*; Springer: New York, NY, USA, 1998.
33. Wu, Z.H.; Li, J.H. Carbon fiber material in musical instrument making. *Mater. Des.* **2016**, *89*, 660–664. [CrossRef]
34. Ono, T.; Takahashi, I.; Takasu, Y.; Miura, Y.; Watanabe, U. Acoustic characteristics of wadaiko (traditional Japanese drum) with wood plastic shell. *Acoust. Sci. Technol.* **2009**, *30*, 410–416. [CrossRef]
35. Phillips, S.; Lessard, L. Application of natural fiber composites to musical instrument top plates. *J. Compos. Mater.* **2012**, *46*, 145–154. [CrossRef]
36. Koruk, H.; Genc, G. Investigation of the acoustic properties of bio luffa fiber and composite materials. *Mater. Lett.* **2015**, *157*, 166–168. [CrossRef]
37. Holzweissig, F.; Leissa, A.W. Vibration of Plates. (Nasa Sp-160). VII + 353 S. m. Fig. Washington 1969. Office of Technology Utilization National Aeronautics and Space Administration. Preis brosch. $ 3.50. Available online: https://onlinelibrary.wiley.com/doi/abs/10.1002/zamm.19710510331 (accessed on 1 September 2019).
38. Leissa, A.W. Vibration of shells. *J. Chem. Inf. Model.* **1973**, *288*, 1689–1699. [CrossRef]
39. Jones, R.M. *Mechanics of Composite Materials*; Taylor & Francis: Philadelphia, PA, USA, 1999.
40. Ibáñez, M.; Gimenez, A.; Sanchez, F. Material Characterization and Vibro-Acoustic Analysis of a Preimpregnated Carbon Fiber Drum Shell. In Proceedings of the 20th International Conference on Composite Materials, Copenhagen, Denmark, 19–24 July 2015.
41. Centea, T.; Grunenfelder, L.K.; Nutt, S.R. A review of out-of-autoclave prepregs—Material properties, process phenomena, and manufacturing considerations. *Compos. Part. A Appl. Sci. Manuf.* **2015**, *70*, 132–154. [CrossRef]
42. Levy, A.; Kratz, J.; Hubert, P. Air evacuation during vacuum bag only prepreg processing of honeycomb sandwich structures: In-plane air extraction prior to cure. *Compos. Part. A Appl. Sci. Manuf.* **2015**, *68*, 365–376. [CrossRef]
43. Kourkoutsaki, T.; Comas-Cardona, S.; Binetruy, C.; Upadhyay, R.K.; Hinterhoelzl, R. The impact of air evacuation on the impregnation time of Out-of-Autoclave prepregs. *Compos. Part. A Appl. Sci. Manuf.* **2015**, *79*, 30–42. [CrossRef]
44. Hamill, L.; Centea, T.; Nutt, S. Surface porosity during vacuum bag-only prepreg processing: Causes and mitigation strategies. *Compos. Part. A Appl. Sci. Manuf.* **2015**, *75*, 1–10. [CrossRef]
45. Saenz-Castillo, D.; Martín, M.I.; Calvo, S.; Rodriguez-Lence, F.; Güemes, A. Effect of processing parameters and void content on mechanical properties and NDI of thermoplastic composites. *Compos. Part. A Appl. Sci. Manuf.* **2019**, *121*, 308–320. [CrossRef]
46. Tai, J.-H.; Kaw, A. Transverse shear modulus of unidirectional composites with voids estimated by the multiple-cells model. *Compos. Part. A Appl. Sci. Manuf.* **2018**, *105*, 310–320. [CrossRef]
47. Turteltaub, S.; de Jong, G. Multiscale modeling of the effect of sub-ply voids on the failure of composite materials. *Int. J. Solids Struct.* **2019**, *165*, 63–74. [CrossRef]
48. Grunenfelder, L.K.; Nutt, S.R. Void formation in composite prepregs—Effect of dissolved moisture. *Compos. Sci. Technol.* **2010**, *70*, 2304–2309. [CrossRef]

49. Zheng, C.; Liang, S. Preparation and damping properties of medium-temperature co-cured phenolic resin matrix composite structures. *Compos. Struct.* **2019**, *217*, 122–129. [CrossRef]
50. Zhang, H.; Ding, X.; Li, H.; Xiong, M. Multi-scale structural topology optimization of free-layer damping structures with damping composite materials. *Compos. Struct.* **2019**, *212*, 609–624. [CrossRef]
51. ISO 6721-3 Plastics-Determination of Dynamic Mechanical Properties—Flexural Vibration—Resonance Curve Method. Available online: https://www.iso.org/standard/13169.html (accessed on 1 September 2019).
52. Graesser, E.J.; Wong, C.R. *ASTM Special Technical Publication*; ASTM: West Conshohocken, PA, USA, 1992; pp. 316–343.

Article

Dynamic Range Compression and the Semantic Descriptor Aggressive

Austin Moore

Centre for Audio and Psychoacoustic Engineering, School of Computing and Engineering, University of Huddersfield, Huddersfield HD1 3DH, UK; a.p.moore@hud.ac.uk

Received: 29 January 2020; Accepted: 12 March 2020; Published: 30 March 2020

Featured Application: The current study will be of interest to designers of professional audio software and hardware devices as it will allow them to design their tools to increase or diminish the sonic character discussed in the paper. In addition, it will benefit professional audio engineers due to its potential to speed up their workflow.

Abstract: In popular music productions, the lead vocal is often the main focus of the mix and engineers will work to impart creative colouration onto this source. This paper conducts listening experiments to test if there is a correlation between perceived distortion and the descriptor "aggressive", which is often used to describe the sonic signature of Universal Audio 1176, a much-used dynamic range compressor in professional music production. The results from this study show compression settings that impart audible distortion are perceived as aggressive by the listener, and there is a strong correlation between the subjective listener scores for distorted and aggressive. Additionally, it was shown there is a strong correlation between compression settings rated with high aggressive scores and the audio feature roughness.

Keywords: dynamic range compression; music production; semantic audio; audio mixing; 1176 compressor; FET compression; listening experiment

---

1. Introduction

1.1. Background

In addition to general dynamic range control, it is common for music producers to use dynamic range compression (DRC) for colouration and non-linear signal processing techniques, specifically to impart distortion onto program material. Scholarly work has researched DRC use and has shown the industry has developed standard practices which mix engineers implement in their work [1,2]. One such standard is the use of Universal Audio 1176 (and, under its original name, Urei 1176) as a vocal compressor, particularly when processing vocals in popular music mixes where a specific character or distortion is desirable [3]. Users describe the sound quality of vocals processed in this manner with a number of subjective descriptors. This article investigates one common descriptor, "aggressive" to determine what it means at an objective level and answer empirically how an aggressive sound quality can be achieved when using the Universal Audio 1176 (abbreviated to simply 1176) and, more broadly, DRC in vocal productions. The findings will be of use to developers of software and hardware compressors, intelligent mixing algorithms [4,5] and industry professionals, as the novel results have the potential to change and speed up their workflow. More broadly, the work carried out in this article fills a gap in the research, as little work has been done to get a better understanding of the perceptual effects of dynamic range compression during the mixing stage of music production.

Some findings reported in this article were originally presented at an Audio Engineering Society conference [6].

The literature relating to compression in music production has focused mainly on the effects of hyper-compression, often concentrating on whether its artefacts are detrimental to the perceived quality of audio material [7–9]. Taylor and Martens [10] claim that achieving loudness is a significant motivation for using compression, particularly in mastering, so one can argue that this is why hyper-compression is well researched.

Ronan et al. [11] investigated the audibility of compression artefacts among professional mastering engineers. For the study, twenty mastering engineers undertook an ABX listening experiment to determine whether they could detect artefacts created by limiting. Two songs were processed using the Massey L2007 digital limiter to achieve −4 dB, −8 dB and −12 dB of gain reduction. The masters (including the uncompressed versions) were then presented to the listeners using the ABX method. The results showed that the mastering engineers found it challenging to discern differences between a number of the audio tracks, particularly those with −8 dB of gain reduction and the unprocessed reference. The same experiment carried out by Ronan et al. [12] used untrained listeners and showed that they were unable to detect up to 12 dB of limiting.

Campbell et al. [13] had participants rate mixes with compression, which had been introduced at various points in the signal chain, namely on discrete tracks, subgroups and the master stereo buss. Their results showed that listeners preferred mixes where compression had been applied to individual tracks rather than to groups or on the stereo buss. However, their test used identical attack and release settings on all stimuli and made use of the same compressor (Pro Tools Compressor/Limiter), which may have played a role in the results. Adjusting the compressor settings so they were more appropriate for mix buss processing, and using a compressor with a suitable character for buss processing could have yielded a different outcome.

Wendl and Lee [14] looked into the effect of perceived loudness when using compression on pink noise split into octave bands. For this study, the authors wanted to observe if playback level and crest factor (a measurement of peak to RMS) affected perceived loudness after varying amounts of limiting had been applied. The results showed that there was a non-linear relationship between the octave bands and changes to the crest factor. Of interest to professionals is the result which shows that modifications to the crest factor in a band centered around 125 Hz does not correlate with perceived loudness at moderate playback levels. Moreover, the perceived loudness could be louder than one might expect. The authors recommended that music engineers should be cautious of compression activities that affect the low end.

Some noteworthy work was conducted by Ronan et al. [15] which sits outside of the hyper compression studies reviewed so far. The authors of this paper investigated the lexicon of words used to describe analogue compression. Ronan et al. conducted a discourse analysis on 51 reviews of analogue mastering compressors to look for common terms used in the texts and created inductive categories to group the words. The categories they created included signal distortion, transient shaping, special dimensions and glue. A qualitative investigation of a similar nature was carried out for the current study, and the results are presented in Chapter 3. Interestingly, a number of the descriptors gathered by both studies are similar, but aggressive was not included in the work by Ronan et al., suggesting that this descriptor is not commonly used to describe the sound of mastering compression.

Other connected work has been carried out by scholars involved with the Semantic Audio Labs and the Semantic Audio Feature Extraction (SAFE) project. SAFE aims to understand the audio features associated with semantic descriptors, which can then be used to create intelligent mixing tools. Stables et al. [16], investigated terms used to describe signal processing on the mix and conducted hierarchical clustering to look for similarities in the terms. The authors presented dendrograms relating to the signal processing techniques, compression, distortion, equalization and reverb. The word aggressive was not included in any dendrogram and, surprisingly, not in the distortion group. The authors do not stipulate which sources the signal processing techniques were applied to. So, the influence of source on the descriptor is not apparent, which is important to consider and will have played a role in the results. Hence, the current study focuses solely on vocal compression.

Bromham et al. [17] looked into compression ballistics (attack and release settings) and how they affected the perception of music mixes in four styles: Rock, Jazz, HipHop and Electronic Dance Music. They asked participants to rate which ballistic setting was the most appropriate for a genre and to select from a list of given words to describe the sound quality of their preferred setting. They discovered that attack played a more significant role on appropriateness than release, with the result applying most strongly to Jazz and Rock. It should be noted that this study made use of a Solid State Logic (SSL) bus compression emulation, which has a much slower attack time than the 1176 compressor used in the current study.

Not directly related to compression, but in the domain of semantic music production, is the work by Fenton and Lee [18], which aims to develop a perceptual model to measure "punch" in music productions. As noted by Fenton and Lee, punch is an attribute that is used by music listeners to describe a sense of power and weight in audio material. Their work uncovered that punch is related to "a short period of significant change in power in a piece of music or performance" as well as dynamic changes to particular frequency bands in the program material. The authors of the work went on to develop a perceptual model of punch for use in a real-time punch meter. The author of the current study advocates the creation of similar meters to measure other perceptual attributes, such as the one in this study, which can be integrated into modern digital audio workstations (DAWS) for use in music production activities.

As can be seen, apart from a small body of work, that there is a gap in the literature relating to the positive effects of compression, particularly during the mixing process. Work by Dewey and Wakefield [19] has shown that compression is one of the most used mixing tools and the present author's experience as a music production academic and professional suggests that audio engineers and scholars are interested in the character of compression. Therefore, the lack of work in this area is surprising.

1.2. Research Aims

Thus, the work in the current study aims to address several pertinent research questions relating to the use of compression during mixing and the semantic nature of its sonic signature. The following research questions are addressed in this article. Firstly, how do professionals describe the sonic signature of the 1176 compressor when processing a range of sources, but most specifically vocals? Secondly, and derived from the results of the first question, what does the subjective descriptor aggressive mean at an objective level?

To answer these questions, three studies were carried out. Initially, a qualitative study was conducted that asked several experienced users to describe the sound quality of the 1176 when compressing a range of audio sources. Based on their responses, they were then asked to rate the appropriateness of commonly used descriptors in a similarity matrix. The results suggested the descriptor aggressive was a synonym for distortion. Thus, the second stage of testing conducted a subjective listening test using the Audio Perceptual Evaluation (APE) method from the Web Audio Evaluation Tool (WAET) [20]. This tested whether listeners rated mixes with vocals compressed by 1176 hardware and using settings measured to have larger amounts of Total Harmonic Distortion (THD) as the most aggressive. Finally, a subsequent listening test was carried out to ascertain whether distortion, timing behaviour or a mixture of both were the most important factors in creating compressed audio perceived to be aggressive. The reason for this test was due to the 1176's reputation as a fast-acting compressor (particularly when working with time constant settings, which will be addressed in Section 4). Therefore, it could be that argued its fast timing creates the aggressive sonic signature, rather than distortion. This was examined in the second listening experiment, which had vocals processed with a clean software compressor (measured with 0% THD) and set to mimic the timing behaviour of the 1176 as well as material compressed with 1176 hardware and measured to have

1.58% THD. The Klanghelm DC8C software compressor [21] was used for this test as it allows user control over a range of design traits that can be used to match the behaviour of several compressors. Most importantly, when used in its clean mode, it does not generate any distortion, even at the fastest time constants.

2. Qualitative Studies

2.1. Professional User Questionaire

An online questionnaire was created using the survey tool Qualtrics [22], which asked experts to describe the sound quality of the 1176 when compressing vocals, drum shells (bass and snare drums), room mics (ambient recordings of the drums in a room) and a bass guitar. Judgement sampling (as opposed to random sampling) was used to select experienced engineers and academics to complete the questionnaire. For an expert to be included they had to be knowledgeable in music production and familiar with the 1176. Judgement sampling does, however, have its limitations and is prone to bias [23]. Thirty-five respondents completed the questionnaire.

The results in Table 1 presents the ten most frequently used descriptors to describe the sound quality of the 1176 compressor for all four sound sources. Aggressive is the most popular word and is investigated in the current study. The other descriptors are likely to refer to amplitude modulation effects (pumping), transient reshaping (punch), colouration (forward, full, midrange and presence) and distortion (dirty and gritty). One would expect fast to be a description of the time constant speed, and not necessarily a description of sound quality.

Table 1. The top ten most frequently used descriptors to describe the sound of the 1176 compression.

Descriptor	Frequency of Occurrence
Aggressive	21
Pumping	11
Forward	10
Punch	8
Full	8
Midrange	8
Fast	7
Presence	7
Dirty	6
Gritty	6

Table 2 shows the most common words used to describe the sound quality of the 1176 when compressing a vocal, which is the main focus of this article. To reduce the number of words in the table, only those recorded more than once have been included. As can be seen, the descriptor aggressive is, again, the most popular, followed by the word gritty, which, as stated previously, is arguably a synonym for distorted. Other descriptors appear to refer to coluration (forward, midrange, presence, full, upfront and sparkly), amplitude modulation (pumping) and potentially the perceptual effect of the attack and release curve (smooth).

Table 2. Descriptors used for 1176 vocal compression.

Descriptor	Frequency of Occurrence
Aggressive	6
Gritty	5
Forward	4
Midrange	4
Presence	4
Full	3
Sparkly	2
Up Front	2
Pumping	2
Smooth	2

2.2. Similarity Matrix

To help clarify the meaning of the descriptors, a second stage asked respondents to rate the appropriateness of the most popular descriptors in describing the sound quality of a given compression technique. The compression techniques were: linear processing, colouration general, colouration frequency related, distortion, modulation/altering rhythmic feel, general dynamic range control, attenuating transient and accentuating transient. The author selected these techniques based on prior research [24] which indicated these practices were commonly used by industry professionals when applying dynamic range compression. Respondents completed the task online and recorded their scores on a similarity matrix. This was conducted by creating an online spreadsheet that had compression activities on the X-axis and descriptors on the Y-axis. Respondents then allocated each descriptor a score between zero and four to rate its appropriateness (zero being totally inappropriate and four being totally appropriate). As a descriptor could relate to more than one compression technique, the respondents were instructed to rate the descriptor for as many techniques as they felt appropriate. The use of similarity matrices to look for associations between audio descriptors has been used in several previous studies [25–28].

The similarity matrix was completed by twelve experts, all of which had participated in the previous stage. The results were averaged to take the mean score for each descriptor and clustering was conducted using the Euclidean distance and Ward methods. The statistical software R was used to generate the dendrogram shown in Figure 1. As shown in the dendrogram, there are several subsets, illustrated in brown, green, blue and grey. Inspection of the grey subset highlights descriptors which appear to fall into two main categories, one relating to dynamics and the other pertaining to colouration and distortion (linear and non-linear distortion). Looking at a lower branch in the dendrogram highlights that aggressive is grouped with, among other words, attitude, energy and smashed. All of these words are subjective and have not been defined in any of the previous literature. Moreover, one could argue these descriptors relate to the character of the compressor and are potentially multi-dimensional attributes. Referring back to the descriptors used in Table 1, this shows that contentions made previously regarding the meaning of these words are generally correct. As an example, the terms dirt and gritty are connected to distortion in the brown subset. Pumping is connected to ambience (which is understandable as compression-related amplitude modulation often manifests as quick changes in the amplitude of the ambience present in the program material). Finally, punch is connected to definition and also attack, which would support the notion that this descriptor is related to the manipulation of transient shapes.

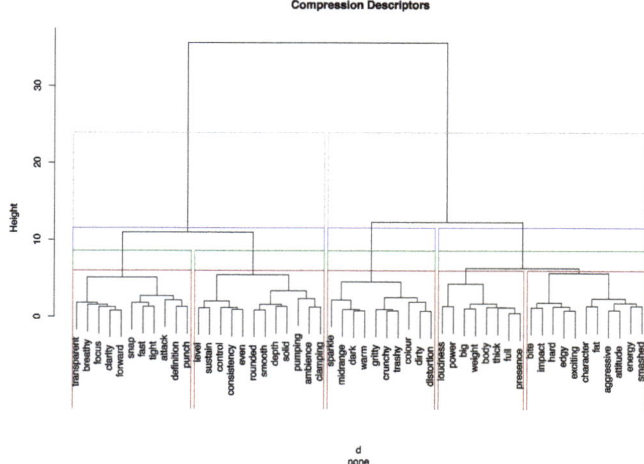

Figure 1. Results of clustering using the Ward method. The descriptors are split into five subsets (brown), four subsets (green), three subsets (blue) and two subsets (grey).

To get a better understanding of the descriptor aggressive, the focus of this study, statistical analysis was conducted on the mean scores allocated by the respondents to the compression techniques in relation to the term aggressive. The results show that there was a statistically significant difference between groups (compression techniques), as determined by one-way ANOVA (F $(7,88)$ = 3.854, p = 0.001). A Tukey post-hoc test revealed that the experts considered the descriptor aggressive was statistically significantly lower for the compression techniques "general dynamic range control" (p = 0.027), "modulation" (p = 0.002) and "linear processing" (p = 0.001) compared to the compression technique "distortion". There was no statistically significant difference between the descriptor aggressive score for the compression techniques "colouration general" (p = 0.229), "colouration frequency related" (p = 0.171), "attenuating transient" (p = 0.088) and "accentuating transient" (p = 0.124) compared to the compression technique "distortion". The reason for the lack of significance between these techniques is thought to be as a result of distortion reshaping the transient portion of the audio material (particularly true for attenuating the transient) and the introduction of harmonic components, which leads to colouration. Therefore, it appears that, from this study, engineers consider the descriptor aggressive to relate to compression techniques that distort and colour the audio and, to a lesser extent, reshape the transient portion of the program material.

3. Preliminary Objective Tests

3.1. Choice of Compressor Time Constant Settings

In preparation for perceptual listening experiments, work was conducted to ascertain appropriate time constant settings for use in the experiments. The 1176 has continuously variable attack and release controls. Thus, a large amount of possible combinations are available. However, it would not be practical to use all of these in listening experiments as a large number of stimuli is known to cause listener fatigue [29]. Thus, content analysis [30] was conducted on 1176 vocal compression settings, created by professional engineers for the 1176 UAD plugin. This work was conducted to discover the most popular settings, which could then be used in the creation of stimuli for the listening experiments. The results revealed that specific combinations of attack and release settings were regularly used, with release positions between five and seven and attack positions between one and three being most common. Additionally, it was noted that the 4:1 ratio was often implemented for general vocal settings and the all-buttons mode (a popular "special mode" achieved by depressing all ratio buttons simultaneously) was employed for highly coloured processing. Table 3 shows how frequently particular settings are used in the vocal presets. The bottom two rows of the table show that positions between one and four are most common for attack and positions between five and seven most common for release. Anecdotal evidence by the author supports this result as they have observed many professional audio engineers setting the 1176 compressor with these time constant settings.

Table 3. Popularity of 1176 time constant settings. The table illustrates how often a setting was used by professionals.

Setting	Release Setting Used	Attack Setting Used
1	0%	46.67%
2	0%	20%
3	0%	20%
4	18.18%	6.67%
5	18.18%	6.67%
6	63.64%	0%
7	0%	0%
1–4	0%	93.33%
5–7	100%	6.67%

Based on these findings, the following attack and release combinations were used in the following listening experiment (attack is abbreviated to A and release is abbreviated to R): A3R7, A1R7, A3R5, A1R5. The combinations were used in both the 4:1 and all-buttons ratio modes. More general research of content on the 1176 [31] showed the A3R7 combination to be a popular setting for a range of instrument sources. Therefore, the settings used in the experiment are considered by the author to be representative of real working scenarios. It is also worth bearing in mind the attack control on the 1176 is quoted as ranging between 20–800 microseconds and critical listening by the author revealed very little difference in sound quality between any attack time between positions one and four. Additionally, the reader should consider that the attack and release controls on the 1176 work counterclockwise, meaning attack and release position seven is the fastest and one the slowest.

3.2. Distortion Characteristics

A series of total harmonic distortion (THD) measurements were made on the 1176 at various attack and release configurations to observe how time constant settings affected the distortion characteristics of the compressor. The measurements for release were made by keeping the attack time fixed at its fastest setting, seven, and making a measurement at each release position. The measurements for attack were made by restricting the release time to its quickest setting, seven, and making measurements at each attack position. During testing, the compressor was adjusted to achieve −10 dB of gain reduction and a 1 kHz test tone was used as the input signal. The output of the compressor was recorded on a Digital Audio Workstation (DAW) at 24 bit/44.1 kHz. The results showed distortion artefacts reduced significantly when using release times slower than position five and that the attack control had a smaller effect on the reduction of distortion. Furthermore, higher ratios had the effect of increasing non-linearity, with the all-buttons mode increasing non-linearity significantly more than any other ratio. Plot (a) in Figure 2 illustrates the effect of lengthening the attack and release time on THD. As can be seen, there is a sharp drop off in THD up to release position five and a small reduction in THD with attack times slower than position seven. Plot (b) in Figure 2 shows the same measurements made in all-buttons, which is a so-called "special" ratio mode afforded by the 1176 compressor. Although not originally intended for use, it was found by music producers that depressing all the ratio buttons simultaneously produced intriguing sonic behaviour by the compressor. In actual terms, the FET's bias is set outside of calibration, resulting in a significant increase in distortion. Note the much larger amount of THD in this setting, but similar drop off in amount as the release and attack speeds are reduced.

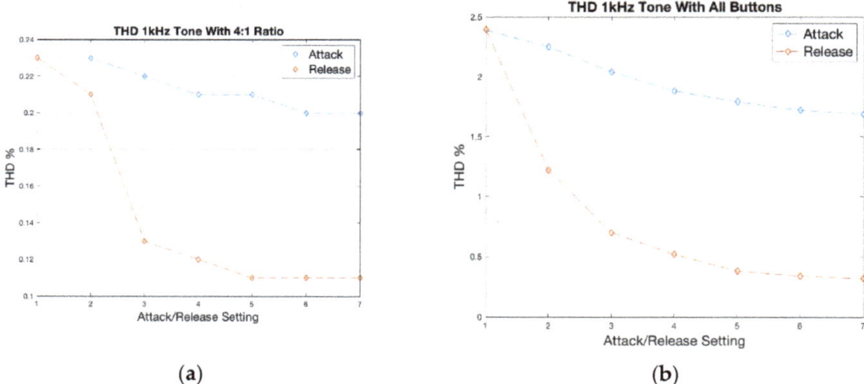

Figure 2. (**a**) Total Harmonic Distortion (THD) as a function of attack and release using a 4:1 compression ratio and a 1 kHz test tone; (**b**) THD as a function of attack and release using the "all-buttons" compression ratio and a 1 kHz test tone.

Figure 3a,b are plots created using Matlab's THD function. They represent an FFT display which illustrates the nature of the harmonic distortion. To create the plots, a 1 kHz tone was input to the compressor to achieve −10 dB of gain reduction, with the time constants set for attack at three and release at seven. The output of the compressor using a 4:1 ratio and the all-buttons mode was then recorded into a DAW at 22 bit/44.1 kHz. The figures make clear the differences in distortion characteristics between the 4:1 ratio and the all-buttons mode. As one might expect, the distortion components are integers of the 1 kHz test tone and consist of a mixture of odd and even order harmonics. In Figure 3b, it is worth noting the significant increase in the amplitude of all components and also an increase in higher-level harmonics. Critical listening to an audio example of the compressed test tone reveals the all-buttons mode is highly distorted with increased brightness as a result of the additional high-level and high-order distortion components.

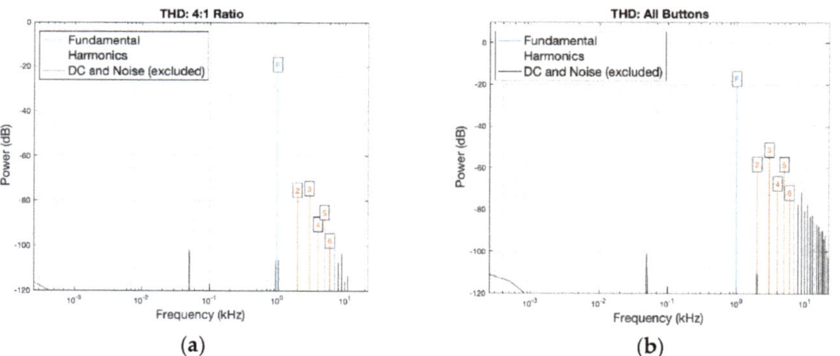

Figure 3. (**a**) Distortion components created using a 4:1 ratio and attack at three and release at seven; (**b**) distortion components created using the all-buttons ratio and attack at three and release at seven.

The attack and release settings, which were shown to be commonly used in Table 3, are the settings that generate the most distortion. Thus, it appears that professional mix engineers are, perhaps albeit unbeknown to themselves, actively seeking out distortion from the 1176 in their workflow. Therefore, the aggressive sound quality once again appears to pertain to distortion. However, additional perceptual testing was required to substantiate this hypothesis. THD measurements made on the settings used in Listening Experiment 1 can be seen in Table 4, where the effect that both the attack and release and the all-buttons mode have on non-linearity can be observed.

Table 4. THD measurements made using a 1 kHz tone and the time constants and ratios used in Listening Experiment 1.

Setting	THD
A3R7 All	1.58%
A1R7 All	1.51%
A3R5 All	0.54%
A1R5 All	0.50%
A3R7 4:1	0.25%
A1R7 4:1	0.24%
A3R5 4:1	0.17%
A1R5 4:1	0.16%

4. Perceptual Listening Experiments

4.1. Listening Experiment 1 Method

To test the hypothesis, a subjective listening test was devised using the Web Audio Evaluation Tool (WEAT), which made use of the Audio Perceptual Evaluation (APE) method. Stimuli were created by processing the vocal from two separate rock songs with, 1176 hardware, using the attack/release combinations mentioned previously. To restrict the number of stimuli, the amount of compression was limited to one setting, which was −10 dB of gain reduction. Ciletti et al. note that to best assess the sonic signature of a compressor, it is advisable to use the device in a heavy state of compression [32]. The amount of gain reduction was measured to show an average of −10 VU on the gain reduction meter. The compressed vocals were then mixed back into the audio tracks, which were level matched to −23 LUFS. In addition, a mix making use of the uncompressed vocal was used to create a total of nine stimuli per song. All audio was recorded and processed at 24 bit/44.1 kHz.

Listeners were presented the stimuli on four separate screens of the listening test (two per song), where they were asked to rate the amount of perceived distortion on two screens and the amount of perceived aggression on the remainder. Scales on the interfaces were labelled from least distorted to most distorted and least aggressive to most aggressive and were measured on a scale from zero to one. Participants were not instructed explicitly what aggressive meant, as the author wished to avoid training the listeners with their interpretation of the descriptor. The order of the audio and screens were randomized to prevent bias, and the test was carried out by 17 expert listeners in a university music laboratory environment using Sennhiesier HD650 headphones on iMac computers. The sample size was considered to be appropriate for a test of this nature and is commensurate with ITU recommendations [33].

4.1.1. Listening Experiment 1 Results and Discussion

The results from the listening test can be seen in Figure 4, where (a) shows the mean result for the descriptor aggressive with a 95% confidence interval for both songs and all time constant settings tested. As can be seen, there is little difference between the time constant settings for both the ratios tested, but there is a difference between the uncompressed material, the 4:1 ratio and the all-buttons mode. It is worth noting that the two all-buttons modes that measured highest for THD (see Table 3 for THD results) are not rated any higher than the other two all-buttons settings. An inspection of the FFT plots suggests this is a result of the even-order harmonics remaining at fairly consistent levels across the four settings, while the odd-order harmonics are attenuated as attack and release are slowed. This results in a lower THD measurement, which evidently does not result in a perceptually less aggressive sonic signature. It should be noted that many of the participants reported that the difference between some of the stimuli was small, and they found the test to be challenging. Therefore, the effect of listener fatigue should be kept in mind. The ratings for distortion are illustrated in Figure 4b where a similar trend is visible. Once again, there is a difference between the ratios and the uncompressed material, but no difference between the different time constant settings for the two ratio settings. Thus, it appears that there is little perceptual difference between the time constant settings used in the current study, but the additional harmonic distortion created in the all-buttons mode is noticeable to the participants of the experiment.

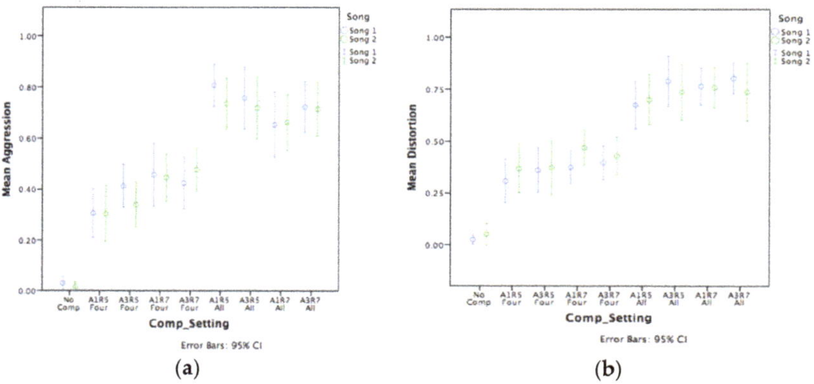

Figure 4. (**a**) Results for the descriptor aggressive from listening experiment 1; (**b**) results for the distortion from listening experiment 1. Note, no significance was found between time constant settings, but significance was found between ratio settings.

Audio features pertaining to the noise-like properties of sound (Roughness and Zero Crossing Rate) were extracted from the vocal tracks using MIRtoolbox for Matlab [34] and are presented in Table 5. The results for roughness show that the feature increases in value between the uncompressed audio and both ratio settings, and also between 4:1 and all-buttons mode. Within the time constant settings for each ratio, the results with the release time set to seven are the highest and this is largely comparable with the THD results shown in Table 4. However, the A1R7 combination for both ratio settings has slightly larger roughness values than the A3R7 combination, while the THD results highlight the A3R7 combination as having the largest amount of THD. The similarity in results within the ratio settings for roughness may be another reason why listeners rated the time constant settings comparably, despite the variation in THD. The values for zero crossing rate (ZCR) are less revealing, with no clear pattern in the results emerging, apart from an increase in ZCR when using compression.

Table 5. Roughness and zero crossing rate (ZCR) features extracted from the vocal material used in test one.

Setting	Roughness		Zero Crossing Rate	
	Song 1	Song 2	Song 1	Song 2
No Comp	33.73	26.84	1887.92	1676.40
A1R5 4:1	99.12	105.58	2909.15	2155.41
A1R7 4:1	129.12	130.46	2915.28	2123.67
A3R5 4:1	98.97	102.17	2579.37	2166.41
A3R7 4:1	128.7	130.65	2484.85	2125.73
A1R5All	202.85	236.18	2966.41	2053.57
A1R7All	212.88	241.84	2850.03	2055.71
A3R5All	199.26	232.14	2881.21	2083.53
A3R7All	209.87	247.17	2953.82	2067.16

4.1.2. Statistical Analysis of Experiment 1 Results

A two-way repeated measures ANOVA was run to determine the effect of compression settings and the interaction effect and compression settings of the two songs on perceived distortion. Mauchly's test of sphericity indicated that the assumptions of sphericity had been violated for the two-way interaction between the song and settings $\chi^2(2) = 73.13$, $p = 0.001$. Therefore, a Greenhouse–Geisser correction was applied ($\varepsilon = 0.580$). Mauchly's test of sphericity indicated that the assumptions of sphericity had not been violated for the effect of the settings $\chi^2(2) = 48.45$, $p = 0.081$.

Simple main effects were run and showed that there was no statistically significant two-way interaction between the songs and settings on perceived distortion., $F(8,128) = 0.648$, $p = 0.653$. There was, however, a statistically significant effect of the settings on perceived distortion, $F(8,128) = 50.97$, $p < 0.001$. Post-hoc analysis with a Bonferroni adjustment showed the mean distortion scores for the 4:1 and all-buttons settings were statistically significantly higher than the scores for no compression ($p < 0.001$). In addition, the mean distortion scores for the all-buttons settings were statistically significantly higher than the scores for the 4:1 ratio settings ($p < 0.001$). Within the four different time constant settings used for both 4:1 and all-buttons, there was no statistical significance.

A second two-way repeated measures ANOVA was run to determine the effect of the compression settings and the interaction and compression settings of the two songs on perceived aggression. Again, Mauchly's test of sphericity indicated that the assumptions of sphericity had been violated for the two-way interaction between the song and settings $\chi^2(2) = 53.99$, $p = 0.028$; thus, a Greenhouse–Geisser correction was applied ($\varepsilon = 0.531$). Mauchly's test of sphericity indicated that the assumptions of sphericity had not been violated for the effect of settings $\chi^2(2) = 27.98$, $p = 0.081$.

Simple main effects were run and showed that there was no statistically significant two-way interaction between the songs and settings on aggressive sound quality, $F(8,128) = 0.301$, $p = 0.886$. There was, however, a statistically significant effect of settings on aggressive sound quality, $F(8,128) = 69.26$, $p < 0.001$, suggesting that settings have a statistically significant effect on aggressive sound quality. Post-hoc analysis with a Bonferroni adjustment showed the same statistical significance between no compression and the two ratio settings as reported previously for distortion. Again, there was no statistical significance within the four different time constant settings used for either 4:1 or all-buttons, meaning that, in the current study, different time constant settings have no significant effect on the perception of distortion or an aggressive sound quality.

The mean scores for aggressive and distortion were analyzed to assess if there was a statistically significant correlation between the scores. Both variables (aggressive and distortion) for both songs were normally distributed, as assessed by a Shapiro–Wilk test ($p > 0.05$); thus, the variables were investigated for correlation. Pearson's product–moment correlation was run to determine the relationship between perceived aggressive and distortion scores and the results show that there is a strong correlation between the mean scores for aggressive and distortion, which is statistically significant for song one ($r = 0.960$, $n = 9$, $p = 0.001$) and song two ($r = 0.983$, $n = 9$, $p = 0.001$). A scatter plot of the mean scores for aggressive and distortion for both songs is illustrated in Figure 5, where the correlation between the two can be clearly observed.

Correlation between the aggressive scores and the roughness features extracted from the vocal files was investigated by running Pearson's product–moment correlation. The results show a strong correlation between roughness and aggressive, which is statistically significant for song one ($r = 0.968$, $n = 9$, $p = 0.001$) and song two ($r = 0.962$, $n = 9$, $p = 0.001$).

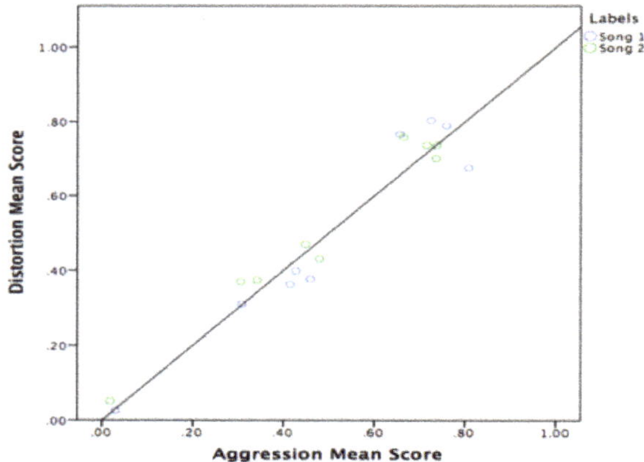

Figure 5. Scatter plot for aggression and distortion mean scores.

4.2. Listening Experiment 2 Method

The previous test demonstrated that vocals compressed with settings measured to have greater than or equal to 0.5% THD were rated as being the most aggressive. However, it could be argued the timing behavior of the 1176, particularly when working in all-buttons mode, plays a role in the result. Therefore, a second test was devised, which aimed to decouple distortion and timing behavior and answer whether distortion, timing behavior or a mixture of both were the key components in the creation of an aggressive sound quality. The experiment made use of the APE listening test interface and had participants rate the vocal tracks of three separate songs on the aggressive quality of the vocals. The two songs used in the previous experiment were utilized again, as well as a third new rock song, which was added to give the results more validity over a wider range of test scenarios. Participants were also asked to comment on the audio they were hearing using up to three descriptors.

During the previous experiment, it was found the time constant settings had no significant effect on an aggressive sound quality; therefore, the vocal tracks were compressed with the hardware 1176 using only the A3R7 time constant (which measured highest for THD) and in 4:1 and all-buttons ratio modes. In addition, the vocals were compressed with the Klanghelm DC8C software compressor, using settings that emulated the timing behaviour of the 1176 in both ratios, and set to measure 0% THD. The timing behaviour was emulated by feeding the hardware 1176 and the software compressor a tone burst and adjusting the parameters of the software compressor until the software closely resembled the timing curve of the 1176 in both settings (see previous work by the author where the tone burst method is used and discussed in detail [31]). While this method did not allow for the exact matching of the 1176's timing curve, it did create very similar results. A more robust method could make use of a specifically designed software compressor algorithm that allows the experimenter to simply turn distortion on and off, but this would require close modelling of the 1176, which was beyond the scope of the current study. Finally, all audio used in Experiment 2 was recorded and processed at 24 bit/44.1 kHz.

4.2.1. Listening Experiment 2 Results and Discussion

The results from the second listening experiment are depicted in Figure 6, which represents the mean result for the descriptor aggressive with a 95% confidence interval for all three songs and all time constant settings tested.

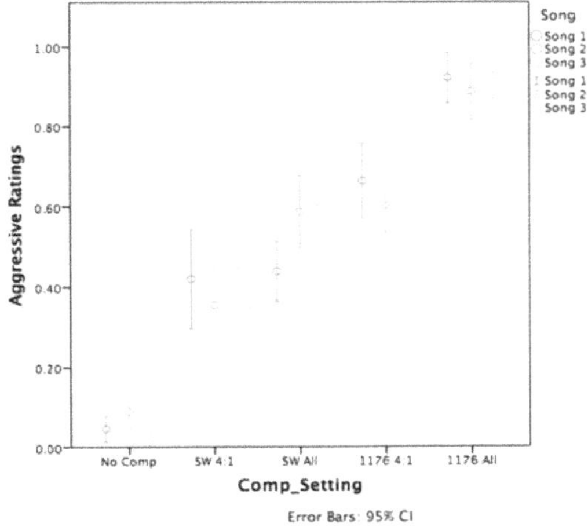

Figure 6. Aggressive results from the second listening experiment.

Looking at the plot, there is an overlap between the scores for SW 4:1 and SW All for songs one and three and an overlap between SW All and 1176 4:1 for song two. However, it is apparent the 1176 all-buttons setting has been rated as the most aggressive for all three songs and the clean software emulation measured to have 0% THD does not score as high as the 1176 all-buttons mode. Thus, the results suggest compression activities that generate audible distortion are needed for the most aggressive vocal sonic signatures.

4.2.2. Statistical Analysis of Experiment 2 Results

A two-way repeated measures ANOVA was run to determine the effect of compression settings measured to have or not have distortion and the interaction effect of the three songs and compression settings on an aggressive sound quality. Mauchly's test of sphericity indicated that the assumptions of sphericity had been violated for the two-way interaction between the song and settings $\chi^2(2) = 71.82$, $p = 0.001$. Therefore, a Greenhouse–Geisser correction was applied ($\varepsilon = 0.578$). Mauchly's test of sphericity indicated that the assumptions of sphericity had not been violated for the effect of the settings $\chi^2(2) = 7.45$, $p = 0.593$.

Simple main effects were run and showed, again, that there was no statistically significant two-way interaction between the songs and settings on aggressive sound quality $F(8,136) = 0.208$, $p = 0.081$. There was, however, a statistically significant effect of settings on an aggressive sound quality, $F(4,68) = 181.722$, $p < 0.001$, suggesting that the settings used have a statistically significant effect. Post-hoc analysis with a Bonferroni adjustment showed the mean aggressive scores for all compressed settings were statistically significantly higher than the scores for no compression ($p < 0.001$). The scores for both the 1176 settings were statistically different from one another ($p < 0.001$), but the scores for both the software settings were not statistically different ($p = 0.57$). This indicates that the faster timing behaviour of the SW All setting, which was emulating the timing curve of the 1176 in all-buttons, has little additional effect over the SW 4:1 setting in the creation of an aggressive vocal sonic signature. The scores for both the 1176 settings were statistically higher than the scores for both the software settings ($p < 0.001$). This indicates that while a clean, fast-acting compressor can give a vocal a more aggressive sound quality than the uncompressed audio, compression settings that impart audible distortion are required for the most significant effect.

4.2.3. Textural Analysis of Descriptors Used by the Participants

Participants of the second listening experiment were encouraged to write descriptors to describe the sound of the vocal in the stimuli they had heard. WAET allows the test designer to include text boxes in the listening test's interface. Thus, participants recorded descriptors into these boxes during listening. A total of 88% of the participants noted descriptors and Figure 7 shows word frequency plots of the twenty most frequently used descriptors for each compressor.

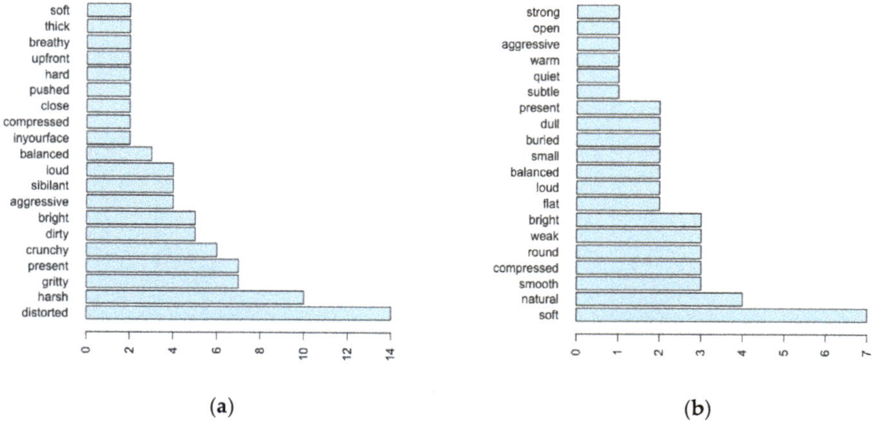

Figure 7. (**a**) Descriptors used for the Universal Audio 1176 (1176) compressor; (**b**) descriptors used for the clean software compressor measured to have 0% THD. Descriptors for both ratios and all three songs have been combined for both compressors.

The word distorted is the most frequently used descriptor for the 1176 compressor. Moreover, descriptors which were shown to relate to distortion in Figure 1, namely gritty, crunchy and dirty also appear often for the 1176. Harsh is also a popular term for this compressor and may be related to distortion. However, one could argue it is a hedonic judgement of preference. Present and bright are two prevalent terms for the 1176, and this is commensurate with the long-term average spectrum (LTAS) plot shown in Figure 8. The LTAS measurements were plotted with a Matlab function [35] using 1/16th octave smoothing. Only one of the songs used in the listening experiments is presented in Figure 8. However, all songs show a similar result, which is that the 1176 has more energy in the high end of the frequency spectrum compared with the uncompressed material and the clean software compressor output. In Figure 8, the increased energy occurs from 4 kHz onwards. Furthermore, the brightness, presence and harshness noted by the participants when listening to 1176 audio, may be related to the descriptor sibilance. Further work should investigate the association between these descriptors by conducting perceptual listening experiments in which the researcher controls these attributes.

Figure 7b illustrates the descriptors used to elucidate the sound quality of the clean software compressor. The most common term is soft, and one can argue that this word is being used as an antonym for aggressive. Natural, smooth, compressed, round, weak and bright are also used by the participants to describe this compressor. Except for bright, they also appear to be terms used to describe the antithesis of an aggressive sound quality. A study carried out by Bernays and Traube [36], which investigated descriptors used to describe piano timbre, found the terms soft, velvety, round and full-bodied were connected. It is worth noting that rounded and smooth are also connected terms in Figure 1. However, further work by the author aims to obtain a better understanding of the most popular descriptors shown in Figure 7 and how they relate to the timbre of DRC.

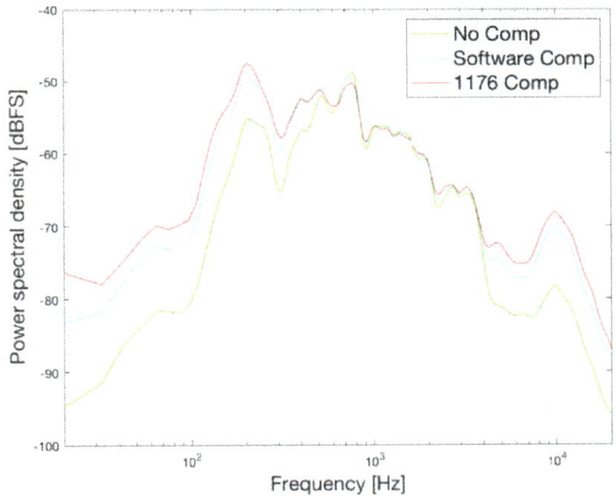

Figure 8. Long-term average spectrum (LTAS) measurement from the uncompressed audio (green), the clean software compressor (blue) and the 1176 compressor in all-buttons mode (red).

5. Conclusions

This paper has shown that professional engineers use the descriptor "aggressive" when describing the sound quality of compression techniques that distort the signal. The first listening experiment demonstrated that there is a strong positive correlation between the listeners' scores for distorted and aggressive when rating the same audio stimuli in a controlled listening experiment. It was also shown that compression settings measured to have 0.5% THD and above were rated as both the most distorted and the most aggressive, but there was no significant difference between settings measured to have more than 0.5% THD. Meaning, in this current study, that listeners could not discern any noticeable difference in perceived distortion or aggression amongst audio measured between 0.5% and 1.58% THD. The various time constant settings used in the experiment, which were gleaned from common settings used in the industry, had no significant effect on the perception of distortion or aggressive sonic signatures. Finally, the experiment revealed a strong correlation between settings rated as aggressive and the audio feature roughness, suggesting that this plays a role in the perception of aggressive sounding audio.

The second listening experiment revealed that compression, which imparts distortion onto the program material, is needed to achieve the most aggressive sound qualities. It appears that fast compression with no distortion (as emulated with the clean software compressor) can affect aggressive sound qualities. Still, the effect is not nearly as significant as using fast-acting compression and distorted artefacts. Both experiments indicated there was no interaction effect between the songs used and the compression settings. Thus, it appears that the songs had little bearing on the results, and the findings from these two experiments should translate to other songs in similar genres.

Finally, a textual analysis conducted on descriptors gathered during the second experiment highlighted the use of descriptors which relate to distortion. The author plans to carry out a new study which will investigate the lexicon of distortion, looking for the similarities between these terms. The results of this study will afford the academic and professional community with a better understanding of how music producers describe and implement distortion in music production.

Funding: This research received no external funding.

Acknowledgments: The author would like to thank Jonathan Wakefield for his help and advice on experimental design.

References

1. De Man, B. *Towards a Better Understanding of Mix Engineering*; Queen Mary University of London: London, UK, 2017.
2. Pestana, P. Automatic Mixing Systems Using Adaptive Audio Effects. Ph.D. Thesis, Universidade Católica Portuguesa, Lisbon, Portugal, 2013.
3. Moore, A. All Buttons in: An investigation into the use of the 1176 FET compressor in popular music production. *J. Art Rec. Prod.* **2012**, *6*.
4. Ma, Z.; De Man, B.; Pestana, P.D.; Black, D.A.; Reiss, J.D. Intelligent multitrack dynamic range compression. *J. Audio Eng. Soc.* **2015**, *63*, 412–426. [CrossRef]
5. Moffat, D.; Sandler, M. Adaptive ballistics control of dynamic range compression for percussive tracks. In Proceedings of the Audio Engineering Society Convention 145, New York, NY, USA, 17–20 October 2018.
6. Moore, A.; Wakefield, J. An Investigation into the Relationship between the Subjective Descriptor Aggressive and the Universal Audio of the 1176 FET Compressor. In Proceedings of the Audio Engineering Society Convention 142, Berlin, Germany, 20–23 May 2017.
7. Nielsen, S.H.; Lund, T. Level control in digital mastering. In Proceedings of the Audio Engineering Society Convention 107, Munich, Germany, 8–11 May 1999.
8. Nielsen, S.H.; Lund, T. 0 dB FS+ Levels in Digital Mastering. In Proceedings of the Audio Engineering Society Convention 109, Los Angeles, CA, USA, 22–25 September 2000.
9. Hjortkjær, J.; Walther-Hansen, M. Perceptual effects of dynamic range compression in popular music recordings. *J. Audio Eng. Soc.* **2014**, *62*, 37–41. [CrossRef]
10. Taylor, R.W.; Martens, W.L. Hyper-compression in music production: Listener preferences on dynamic range reduction. In Proceedings of the Audio Engineering Society Convention 136, Berlin, Germany, 26–29 April 2014.
11. Ronan, M.; Ward, N.; Sazdov, R.; Lee, H. The Perception of Hyper-Compression by Mastering Engineers. *J. Audio Eng. Soc.* **2017**, *65*, 613–621.
12. Ronan, M.; Ward, N.; Sazdov, R. The Perception of Hyper-Compression by Untrained Listeners. In Proceedings of the Audio Engineering Society Conference: 60th International Conference: DREAMS (Dereverberation and Reverberation of Audio, Music, and Speech), Leuven, Belgium, 3–5 February 2016.
13. Campbell, W.; Paterson, J.; van der Linde, I. Listener Preferences for Alternative Dynamic-Range-Compressed Audio Configurations. *J. Audio Eng. Soc.* **2017**, *65*, 540–551. [CrossRef]
14. Wendl, M.; Lee, H. The Effect of Dynamic Range Compression on Perceived Loudness for Octave Bands of Pink Noise in Relation to Crest Factor. In Proceedings of the Audio Engineering Society Convention 138, Warsaw, Poland, 7–10 May 2015.
15. Ronan, M.; Ward, N.; Sazdov, R. Investigating the Sound Quality Lexicon of Analogue Compression Using Category Analysis. In Proceedings of the Audio Engineering Society Convention 138, Warsaw, Poland, 7–10 May 2015.
16. Stables, R.; De Man, B.; Enderby, S.; Reiss, J.D.; Fazekas, G.; Wilmering, T. Semantic description of timbral transformations in music production. In Proceedings of the 24th ACM international conference on Multimedia, Amsterdam, The Netherlands, 15–19 October 2016; pp. 337–341.
17. Bromham, G.; Moffat, D.; Barthet, M.; Fazekas, G. The impact of compressor ballistics on the perceived style of music. In Proceedings of the Audio Engineering Society Convention 145, New York, NY, USA, 17–20 October 2018.
18. Fenton, S.; Lee, H. A Perceptual Model of "Punch" Based on Weighted Transient Loudness. *J. Audio Eng. Soc.* **2019**, *67*, 429–439. [CrossRef]
19. Dewey, C.; Wakefield, J. Elicitation and Quantitative Analysis of User Requirements for Audio Mixing Interface. In Proceedings of the Audio Engineering Society Convention 144, New York, NY, USA, 17–20 October 2018.
20. Jillings, N.; Man, B.D.; Moffat, D.; Reiss, J.D. *Web Audio Evaluation Tool: A Browser-Based Listening Test Environment*; Queen Mary University of London: London, UK, 2015.
21. DC8C Overview. Available online: https://klanghelm.com/contents/products/DC8C/DC8C.php (accessed on 10 January 2020).

22. Snow, J.; Mann, M. *Qualtrics Survey Software: Handbook for Research Professionals*; Qualtrics Labs, Inc.: Provo, UT, USA, 2013.

23. Ponemon, L.A.; Wendell, J.P. Judgmental versus random sampling in auditing: An experimental investigation. *Auditing* **1995**, *14*, 17.

24. Moore, A. *An Investigation into Non-Linear Sonic Signatures with a Focus on Dynamic Range Compression and the 1176 Fet Compressor*; University of Huddersfield: Huddersfield, UK, 2017.

25. Woodcock, J.S.; Davies, W.J.; Cox, T.J.; Melchior, F. Categorization of broadcast audio objects in complex auditory scenes. *J. Audio Eng. Soc.* **2016**, *64*, 380–394. [CrossRef]

26. Berg, J.; Rumsey, F. Verification and Correlation of Attributes Used for Describing the Spatial Quality of Reproduced Sound. 2001. Available online: http://epubs.surrey.ac.uk/542/1/fulltext.pdf (accessed on 10 January 2020).

27. Simurra Sr, I.; Queiroz, M. Pilot experiment on verbal attributes classification of orchestral timbres. In Proceedings of the Audio Engineering Society Convention 143, Berlin, Germany, 20–23 May 2017.

28. Pearce, A.; Brookes, T.; Dewhirst, M.; Mason, R. Eliciting the most prominent perceived differences between microphones. *J. Acoust. Soc. Am.* **2016**, *139*, 2970–2981. [CrossRef] [PubMed]

29. Zielinski, S.; Rumsey, F.; Bech, S. On Some Biases Encountered in Modern Audio Quality Listening Tests—A Review. *J. Audio Eng. Soc.* **2008**, *56*, 427–451.

30. Bos, W.; Tarnai, C. Content analysis in empirical social research. *Int. J. Educ. Res.* **1999**, *31*, 659–671. [CrossRef]

31. Waves Audio CLA Live at Mix LA Part 1/2—YouTube. Available online: https://www.youtube.com/watch?v=7heuq2lV3h4 (accessed on 10 January 2020).

32. Ciletti, E.; Hill, D.; Wolff, P. Gain Control Devices, Side Chains, Audio Amplifiers. Available online: https://www.tangible-technology.com/dynamics/comp_lim_ec_dh_pw2.html (accessed on 10 January 2020).

33. Union, I.T. *Recommendation ITU-R BS. 1116-1*; International Telecommunication Union: Geneva, Switzerland, 1997.

34. Lartillot, O.; Toiviainen, P.; Eerola, T. *MIRtoolbox 1.1 User's Manual*; Finnish Center of Excellence in interdisciplinary Music Research, University of Jyvaskyla: Jyvaskyla, Finland, 2008.

35. Hummersone, C. Long Term Average Spectrum—File Exchange—MATLAB Central. Available online: https://uk.mathworks.com/matlabcentral/fileexchange/55212-long-term-average-spectrum (accessed on 10 January 2020).

36. Bernays, M.; Traube, C. Verbal expression of piano timbre: Multidimensional semantic space of adjectival descriptors. In Proceedings of the Proceedings of the International Symposium on Performance Science (ISPS2011), Toronto, ON, Canada, 24–27 August 2011; European Association of Conservatoires (AEC): Utrecht, The Netherlands, 2011; pp. 299–304.

Article

Design and Application of the BiVib Audio-Tactile Piano Sample Library

Stefano Papetti [1,†], Federico Avanzini [2,†] and Federico Fontana [3,*,†]

1   Institute for Computer Music and Sound Technology (ICST), Zurich University of the Arts,
    CH-8005 Zurich, Switzerland; stefano.papetti@zhdk.ch
2   LIM, Department of Computer Science, University of Milan, I-20133 Milano, Italy;
    federico.avanzini@unimi.it
3   HCI Lab, Department of Mathematics, Computer Science and Physics, University of Udine,
    I-33100 Udine, Italy
*   Correspondence: federico.fontana@uniud.it; Tel.: +39-0432-558-432
†   These authors contributed equally to this work.

Received: 11 January 2019; Accepted: 26 February 2019; Published: 4 March 2019

Abstract: A library of piano samples composed of binaural recordings and keyboard vibrations has been built, with the aim of sharing accurate data that in recent years have successfully advanced the knowledge on several aspects about the musical keyboard and its multimodal feedback to the performer. All samples were recorded using calibrated measurement equipment on two Yamaha Disklavier pianos, one grand and one upright model. This paper documents the sample acquisition procedure, with related calibration data. Then, for sound and vibration analysis, it is shown how physical quantities such as sound intensity and vibration acceleration can be inferred from the recorded samples. Finally, the paper describes how the samples can be used to correctly reproduce binaural sound and keyboard vibrations. The library has potential to support experimental research about the psycho-physical, cognitive and experiential effects caused by the keyboard's multimodal feedback in musicians and other users, or, outside the laboratory, to enable an immersive personal piano performance.

Keywords: musical haptics; piano; auditory feedback; tactile feedback; binaural audio; keyboard vibrations; measurement; recording

---

1. Introduction

During instrumental performance musicians are exposed to auditory, visual and also somatosensory cues. This multisensory experience has been studied since long [1–5], however the specific interaction between sound and vibrations has been object of systematic research since the 1980's [6–12], when tactile and force feedback cues started to be recognized to have a prominent role in the complex perception-action mechanisms occurring during musical instrument playing [13]. More recently, research on the somatosensory perception of musical instruments has been consolidated, as testified by the emerging "musical haptics" topic [14].

This increased interest is partly due to the increased availability of accurate yet affordable sensors and actuators, capable of recording touch gestures and rendering vibrations and force, respectively. Using these devices, complex experimental settings can be realized to measure and deliver multisensory information in a musical instrument, often as a result of a real-time analysis and synthesis process taking place during the performance [15–17]. Once integrated with traditional audio microphone and loudspeaker systems, touch sensors and actuators can be employed first to investigate the joint role of the auditory and somatosensory modality in the perception of a musical instrument, and then to realize novel musical interfaces and instruments building on the lessons of the previous investigation. Through this process, richer or even unconventional feedback cues can be conveyed to the performer, with the aim of increasing engagement, and hence the initial acceptability and subsequent playability of the new instrument [18–21].

In this scenario, the availability of multimodal databases combining and synchronizing different streams of information (audio, video, kinematic data of the instrument and performer in action, physiological signals, interactions among musicians etc.) is increasingly recognized as an essential asset for studying music performance. Recent examples include the "multimodal string quartet performance dataset" (QUARTET) [22], the "University of Rochester multimodal music performance dataset" (URMP) [23], the "database for emotion analysis using physiological signals" (DEAP) [24], the "TU-Note violin sample library" [25]. Furthermore, initiatives aiming at systematizing the creation of these databases have recently appeared, such as RepoVizz [26], a framework for storing, browsing, and visualizing synchronous multimodal data. Overall, this scenario suggests an increasing attention to the design of musical instrument sound databases, motivated by the concrete possibility for their content to be reproduced in instrumental settings including not only loudspeakers and headphones, but also haptic and robotic devices.

The piano represents an especially relevant case study not only for its importance in the history of Western musical tradition, but also for its potential in the musical instruments market due to the universality of the keyboard interface, a feature that has traditionally induced novel musical instrument makers to propose conservative instances of the standard piano keyboard whenever this interface made possible to control even revolutionary sound synthesis methods [27]. It is no accident that, also in recent years, sales of digital pianos and keyboard synthesizers have shown a growing trend as opposed to other instrument sales (https://www.namm.org/membership/global-report). For the same reason, researchers on new musical instruments have steadily elected the piano keyboard as the platform of choice for designing expansions of the traditional paradigm, affording a performer to accurately play two local selections of 88 available tones with the desired amplitude and temporal development [28,29].

When playing an acoustic piano, the performer is exposed to a variety of auditory, visual, somatosensory, and vibrotactile cues that combine and integrate to shape the pianist's perception–action loop. The present authors are involved in a long-term research collaboration around this topic, with a focus on two main aspects. The first one is the tactile feedback produced by keyboard vibrations that reach the pianist's fingers after keystrokes and remains active until key release. The second one is the auditory spatial information in the sound field radiated by the instrument at the performer's head position. Binaural piano tones are offered by a few audio plugin developers (e.g., Modartt Pianoteq (https://www.pianoteq.com/)) and digital piano manufacturers (e.g., Yamaha Clavinova (https://europe.yamaha.com/en/products/musical_instruments/pianos/clavinova/)), but they have limited flexibility of use. Free binaural piano samples can be found, too, such as the "binaural upright piano" library, (https://www.michaelpichermusic.com/binaural-upright-piano) which however offers only three dynamic levels (as opposed to the ten levels provided by the present dataset). More generally concerning the presentation of audio-tactile piano tones, the existing literature is scarce and provides mixed if not contradictory results about the actual perceptibility and possible relevance of this multisensory information [8]. Specific discussions of these aspects have been provided in previously published studies, regarding both sound localization [30] and vibration perception [12] on the acoustic piano. As a notable result of these studies, a digital piano prototype was developed that reproduces various types of vibrations [20] including those generated by acoustic pianos as an unavoidable by-product of their sound production.

Across the years, this research has resulted in the production of an extensive amount of experimental data, most of which resulting from highly accurate measurements with calibrated devices. In an effort to provide public access to such data, the authors of this paper present a dataset of audio-tactile piano samples organized as libraries of synchronized binaural sound and vibration signals.

The dataset contains samples relative to all 88 tones, played at different dynamics on two instruments: a grand and an upright piano. A preliminary version was presented at a recent conference [31], and has now been updated by a new release containing additional upright piano sounds along with the binaural impulse responses of the room in which the same piano was recorded. In order to use this dataset, it is necessary to take into account the recording conditions and, on the user side, to take control of the rendering system in an effort to match the reproduction to the characteristics of the original recorded signals. For this reason, Section 2 describes the hardware/software recording setup and the organization of samples into libraries for the free version of a popular music software sampler. An explanation about how to reproduce the database is provided in Section 3. Section 4 suggests some applications of this library, based on the authors' past design experiences with multimodal piano samples, and on more experiences that are foreseen for future research.

Among past experiences, the most prominent have consisted of studies on the role of haptic feedback during the performance, both vibratory and somatosensory: the former concerns the perception of keyboard vibrations, their accurate reproduction and their effects on the performance [12,20]; the latter concerns the influence of actively playing on the keyboard in the auditory localization process of piano tones [30,32]. Such experiences let to the decision of sharing the dataset with the scientific community, with the goal of fostering the research on the role of vibrations and tone localization in the pianist's perceived instrument quality (both not completely understood yet), as well as adding knowledge about the importance at cognitive level of multisensory feedback for its use in the design of novel keyboard interfaces.

2. Creation of the BiVib library

The BiVib (binaural and vibratory) sample library is a collection of high-resolution audio files (.wav format, 24-bit @ 96 kHz) containing binaural piano sounds and keyboard vibrations, coming along with documentation and project files for its reproduction through a free music software sampler. The dataset, whose core structure is illustrated in Table 1, is made available through an open-access data repository (https://zenodo.org/record/2573232) and released under a Creative Commons (CC BY-NC-SA 4.0) license.

Table 1. Binaural and vibratory (BiVib) core structure. Piano lid configurations are given in square brackets.

	Disklavier Grand	Disklavier Upright
Sample sets (.wav files)	Binaural (closed) Binaural (open) Binaural (removed) Keyboard vibration	Binaural (closed) Binaural (semi-open) Binaural (open) Keyboard vibration
Sampler projects (Kontakt *multis*)	Binaural (closed) + vibration Binaural (open) + vibration Binaural (removed) + vibration	Binaural (closed) + vibration Binaural (semi-open) + vibration Binaural (open) + vibration

2.1. Recording Hardware

The samples were recorded from two Yamaha Disklavier pianos—a grand model DC3 M4 located in Padova (PD), Italy, and an upright model DU1A with control unit DKC-850 located in Zurich (ZH), Switzerland. Disklaviers are Musical Instrument Digital Interface (MIDI from here on)-compliant acoustic pianos equipped with sensors for recording keystrokes and pedaling, and operating electromechanical motors for playback. The grand piano is located in a large laboratory space (approximately 6 × 4 m), while the upright piano is in an acoustically treated small room (approximately 4 × 2 m).

Binaural audio recordings made use of dummy heads for acoustic measurements, with slightly different setups in PD and ZH: the grand piano was recorded with a KEMAR 45BM (GRAS Sound & Vibration A/S, Holte, Denmark), whereas the upright piano with a Neumann KU 100 (Georg Neumann GmbH, Berlin, Germany). Both mannequins were placed in front of the pianos, approximately where the pianist's head is located on average (see Figure 1).

Figure 1. Binaural audio recording setup in Zurich (ZH). Piano lid in 'semi-open' position.

The dummy heads were connected to the microphone inputs of two professional audio interfaces, a RME Fireface 800 (PD, gain set to +40 dB) (Audio AG, Haimhausen, Germany) and a RME UCX (ZH, gain set to +20 dB) (Audio AG, Haimhausen, Germany). The pair of condenser microphones inside the dummy heads were respectively driven by two 26CB preamplifiers supplied by a 12AL power module (PD), and by 48V phantom power provided by the audio interface (ZH).

Three configurations of the keyboard lid were selected for each piano. The grand piano (PD) was measured with the lid closed, fully open, and removed (i.e., detached from the instrument). The upright piano was recorded with the lid closed, in semi-open position (see Figure 1), and fully open. Different lid configurations in fact add insight on the role of the mechanical noise coming from the moving keys to the creation of transient cues of lateralization in the sound field reaching the performer's ears [30]. As a result, three sets of binaural samples were recorded for both pianos—one set for each lid position.

Vibration recordings were acquired with a Wilcoxon Research 736 (Wilcoxon Sensing Technologies Inc., Amphenol, MD, USA) piezoelectric accelerometer connected to a Wilcoxon Research iT100M (Wilcoxon Sensing Technologies Inc., Amphenol, MD, USA) intelligent transmitter, whose AC-coupled output fed one line input of the audio interface. The accelerometer was manually attached with double-sided adhesive tape to each key in sequence, as shown in Figure 2. Its center was positioned 2 cm far from the key edge, where most vibration modes are radiated efficiently, and where pianists typically put their fingertip.

Figure 2. Vibration recording setup. A Wilcoxon Research 736 accelerometer is attached with adhesive tape to a key that is being played remotely via MIDI control.

2.2. Recording Software

Ten values of dynamics were chosen between MIDI key velocity 12 and 111 by evenly splitting this range into eleven intervals. This choice was motivated by a previous study by the present authors, which reported that both Disklaviers produced inconsistent dynamics outside this velocity range [12]. In general, the servo-mechanics of computer-controlled pianos fall short (to an extent that depends on the model) of providing a reliable response at extreme dynamic values [33]. Holding this assumption, two different software setups were used respectively for sampling sounds and vibrations.

Binaural samples were recorded via an automatic procedure programmed in SuperCollider. (A programming environment for sound processing and algorithmic composition: http://supercollider. github.io/.) The recording sessions took place at night time, thus minimizing unwanted noise coming from human activity in the building. On the grand piano, note lengths were determined algorithmically depending on their dynamics and pitch, ranging from 30 s when A0 was played at key velocity 111, to 10 s when C8 was played at key velocity 12. In fact, notes of increasing pitch and/or decreasing dynamics have shorter decay times. These durations allow each note to fade out completely, while minimizing silent recordings and the overall duration of the recording sessions—still, each session lasted approximately six hours.

On the contrary, an undocumented protection mechanism on the upright piano prevents its electromechanical system from holding down the keys for more than about 17 s, thus making a complete decay impossible for some notes, especially at low pitches and high dynamics. In this case, for the sake of simplicity all tones were recorded for just as long as possible.

Vibration samples were recorded through a less sophisticated procedure. A digital audio workstation (DAW) software was used to play back all notes in sequence at the same ten MIDI velocity values as those used for the binaural audio recordings, using a constant duration of 16 s. This duration in fact is certainly greater than the time taken by any key vibration to decay below sensitivity thresholds [12,34].

2.3. Sample Pre-Processing

Because of the mechanics of piano keyboards and the intrinsic limitations of computer-controlled electro-mechanical actuation, a systematic delay is introduced while reproducing MIDI note ON messages, which mainly varies with key dynamics. For this reason, all recorded samples started with silence of varying durations, which had to be removed in view of their use in a sampler (see Section 2.4). Given the number of files that had to be pre-processed (880 for each set), an automated procedure was implemented in SuperCollider to cut the initial silence of each audio sample. The procedure analyzed the amplitude envelope, detected the position of the largest peak, and finally applied a short fade-in starting a few milliseconds before the peak.

Additionally, vibration signals presented abrupt onsets in the first 200–250 ms right after the starting silence, as a consequence of the initial fly of the key and of the following impact against the piano keybed (see Figure 3). These onsets are not related to keyboard vibrations, and therefore they had to be removed.

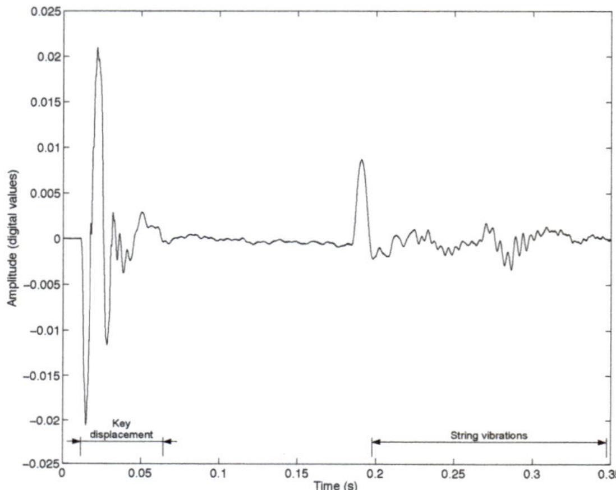

Figure 3. Vibration signal recorded on the grand Disklavier by playing the note A2 at MIDI velocity 12. Picture from [12].

As such onset profiles showed large variability, in spite of several tests made in MATLAB, no reliable automated procedure could be realized for editing the vibration samples. A manual approach was employed instead: files were imported in a sound editor, their waveform was zoomed in and played back, and the onset was cut off.

2.4. Sample Library Organization

The sample library was organized for playback with the 'Kontakt Player' software—a free version of Native Instruments' Kontakt sampler, (https://www.native-instruments.com/en/products/komplete/samplers/kontakt-5-player/) available for Windows and Mac OS systems. The full version of Kontakt 5 was instead used to develop Kontakt project files as described below. The resulting library is organized into several folders, named 'Instruments', 'Multis', 'Resources', and 'Samples'.

The 'Samples' folder—whose total size amounts to about 65 GB—held separate subfolders for the binaural and vibration samples, respectively, which contain further subfolders for each sample set (see Table 1), for example 'grand-open' under the 'binaural' folder.

Independent of their type, sample files were named according to the following mask:

```
[note][octave #]_[lower MIDI velocity]_[upper MIDI velocity].wav,
```

where [note] follows the English note-naming convention, [octave #] ranges from 0 to 8, [lower MIDI velocity] equals the MIDI velocity (range 12–111) used during recording and is the smallest velocity value mapped to that sample in Kontakt (see below), [upper MIDI velocity] is the largest velocity value mapped to that sample in Kontakt. As an example, the file A4_100_110.wav corresponds to the note A from the 4th octave (fundamental frequency 440 Hz) recorded at MIDI velocity 100, and mapped to the velocity range 100–110 in Kontakt. Since the lowest recorded velocity value was 12, for consistency no samples were mapped to the velocity range 1–11 in Kontakt.

Following the terminology used in Kontakt, each instrument reproduces a sample set (e.g., binaural recording of the grand piano with lid open), while each multi combines two instruments respectively reproducing one binaural and one vibration sample set belonging to the same piano. The two instruments in each multi are configured so as to receive MIDI input data on channel 1, thus playing back at once, while their respective outputs are routed to different virtual channels in Kontakt: binaural samples are routed to a pair of stereo channels (numbered 1–2), while vibration samples are played through a mono channel (numbered 3). In this way, when using audio interfaces offering more than two physical outputs, it is possible to render synchronized binaural and vibrotactile cues at the same time by routing the audio signal to headphones and, in parallel, the vibration signal to an amplifier driving one or more actuators.

In each instrument, sample mapping was implemented relying on the 'auto-map' feature found in the full version of Kontakt: this parses file names and uses the recognized tokens for assigning samples to e.g., a pitch and velocity range. The chosen file naming template made it straightforward to batch-import the samples.

The amplitude of the recorded signals was not altered, that is, no dynamic processing or amplitude normalization was applied. The volume of all Kontakt instruments was set to 0 dB. Because of this setting and of the adopted velocity mapping strategy, sample playback was kept as transparent as possible for simplifying the setup of acoustic and vibratory analysis procedures, experiments and interactive applications (see Sections 3 and 4).

3. Application of the BiVib Library

Binaural piano tones such as those offered by Yamaha digital pianos or the Modartt Pianoteq software synthesizer are not fully suitable for research purposes due to being undocumented, hence non reproducible, acquisition procedures and/or post-processing of the sound signals. Moreover, the present authors have no evidence of public samples including piano keyboard vibrations. Thus, BiVib fills two gaps found in the datasets currently available for the reproduction of piano feedback.

3.1. Sample Reproduction

Experiments and applications requiring the use of calibrated data need exact reconstruction of the measured signals at the reproduction side: acoustic pressure for binaural sounds, and acceleration for keyboard vibrations. Obviously the reproduction must take place on a set-up in which neither autonomous sounds nor vibrations are present. For instance, the reproduction of vibrations could take place on a weighted MIDI keyboard (such as those found in digital pianos), while binaural sounds may be rendered through headphones. Figure 4 (left) shows one such setting, in which a commercial digital piano is augmented through the setup schematized as in Figure 4 (right). Note that the bottom of the digital piano has been reinforced by substituting the keybed with a thicker wooden panel, to form what we will call a *haptic* digital piano from here.

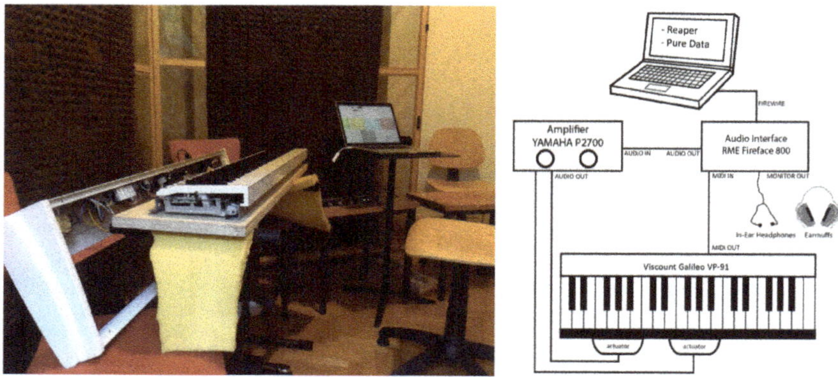

Figure 4. Left: haptic digital piano customization for use with binaural and vibratory (BiVib). **Right**: schematic of a possible reproduction setting. Pictures from [20].

Knowledge of the recording equipment's nominal specifications enable this reconstruction. Such specifications are summarized in a companion document in the 'Documentation' folder. Two examples are listed below:

- Vibration accelerations in m/s^2 are calculated from the vibration samples by making use of the nominal sensitivity values of the audio interface and accelerometer: voltage values are reconstructed from the recorded digital signals (represented by values between -1 and 1) by means of the line input sensitivity of the audio interface, whose full scale level (0 dBFS) was set to $+19$ dBu (reference voltage 0.775 V) for recording. Finally, voltage values are transformed into proportional acceleration values through the sensitivity constant of the accelerometer, equal to 10.2 mV/m/s^2 at 25 °C, with a flat ($\pm 5\%$) frequency response in the range 10–15,000 Hz.

As an example, the digital value 0.001 is equivalent to:

$$20 \times \log_{10}(|0.001|) = -60\,\text{dBFS} \tag{1}$$

and the corresponding voltage value is calculated based on the 0 dBFS level ($+19$ dBu, with 0.775 V reference) as follows:

$$0.775 \times 10^{((19-60)/20)} = 6.9\,\text{mV}. \tag{2}$$

Finally, the matching acceleration value is obtained as:

$$\frac{6.9\,\text{mV}}{10.2\,\text{mV/m/s}^2} = 0.677\,\text{m/s}^2. \tag{3}$$

- Analogously, acoustic pressure values in Pa are obtained from the binaural samples by making use of the relevant nominal specification values of the microphone input chain (microphone, pre-amplifier if present, audio interface input). For the upright piano: microphone sensitivity of 20 mV/Pa, audio interface input gain set to $+26$ dB for an equivalent 0 dBFS level of -16 dBu (reference 0.775 V). For the grand piano: microphone sensitivity 50 mV/Pa, preamplifier gain -0.35 dB, and audio interface input gain set to $+40$ dB for an equivalent 0 dBFS level of -19 dBu (reference 0.775 V).

As an example, the digital value 0.001, equivalent to -60 dBFS as seen in Equation (1), translates to the following voltages:

$$0.775 \times 10^{((-16-60)/20)} = 0.12\,\text{mV} \tag{4}$$

for the upright piano, or

$$0.775 \times 10^{((-19-60+0.35)/20)} = 0.09 \, \text{mV} \tag{5}$$

for the grand piano. The equivalent acoustic pressures are then computed as:

$$\frac{0.12 \, \text{mV}}{20 \, \text{mV/Pa}} = 0.006 \, \text{Pa} \ (\text{about } 49.5 \, \text{dB SPL}) \tag{6}$$

for the upright piano, or

$$\frac{0.09 \, \text{mV}}{50 \, \text{mV/Pa}} = 0.0018 \, \text{Pa} \ (\text{about } 39 \, \text{dB SPL}) \tag{7}$$

for the grand piano.

Once the original measurements have been reconstructed, physical quantities are, in principle, ready for presentation of sounds through headphones and vibrations through tactile actuators, holding the issues and limitations that are listed in the next section.

3.2. Sample Equalization and Limitations

BiVib users should hear the same sounds as the binaural pressure signals measured by the dummy head microphones, and should receive vibrations at the fingers which are identical to the acceleration signals measured by the accelerometer.

The audio reproduction condition can be satisfied by using headphones or earphones that bypass the outer ear and provide a frequency response that is as flat as possible. In this way, the performer would listen to stereo sounds that contain the binaural information created by the dummy head. Note that these sounds also contain the contribution of room reverberation, as the pianos could not be located inside anechoic rooms during the recording sessions: especially the laboratory hosting the grand piano was moderately reverberant, and no reverberation data could be collected for this room at the time of the acquisition of samples. Conversely, the upright piano (ZH) benefited of a silent studio room whose impulse responses have been measured and recently included in BiVib (folder 'IR'). More precisely, responses from two source points generated by a pair of Genelec 8040A loudspeakers were taken in correspondence of the two ear canal entrances of the Neumann KU 100 dummy head with both outer ears removed: As shown in Figure 5, the loudspeakers were placed symmetrically at both sides of the upright piano, 80 cm above the floor, pointing toward the dummy head (angle with the vertical plane parallel to the piano equal to 47°) which was positioned 55 cm far from the upright panel of the piano. Logarithmic sweeps were synthesized in the Audacity audio editor using the Aurora modules [35], then reproduced using the loudspeakers, and finally deconvolved again with Aurora to find four room transfer functions forming the source-ear transfer matrix. Such transfer functions were included in the dataset and can be used to remove the echoes of the recording room through inversion of the transfer matrix, using standard deconvolution techniques [36]. Also, because of the low energy of these echoes, binaural recordings were intentionally left unprocessed, thus allowing musicians and performers to use them as they are, while leaving any decision on their possible manipulation to advanced users.

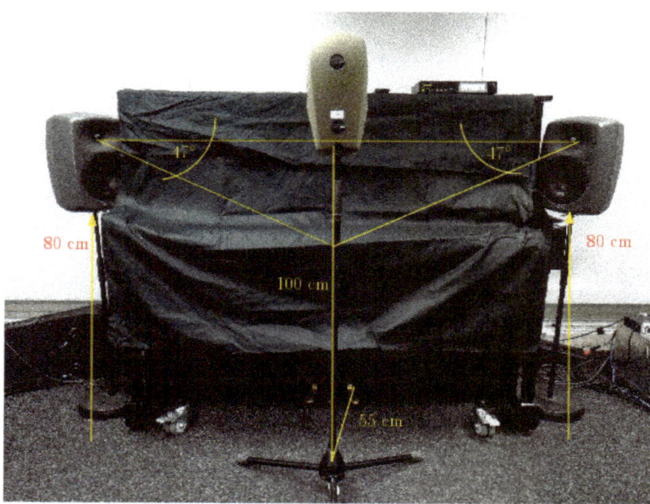

Figure 5. Upright piano recording setup.

The vibration reproduction condition may be ideally satisfied by implementing a weighted MIDI keyboard (such as those found on digital pianos) where each key is provided with an actuator, similar to the prototype technology described in [37]. Even in the unlikely case that such actuators would offer a flat frequency response over the range of interest for tactile sensation [38], vibration reproduction would be altered by the shape, material and construction of the keyboard. That requires the implementation of a compensation procedure—similar to that described in the previous paragraph to remove room acoustic components—that would enable the accurate reproduction of the vibrations recorded by the accelerometer on each key. Unfortunately, on such a keyboard there would be 88 touch points requiring deconvolution. In alternative, two arrays of 88 transfer characteristics can be formed each by measuring vibrations on the same touch points, after exciting the digital piano body with a corresponding pair of tactile transducers.

Figure 6 (left) [12] shows a detail of the haptic digital piano that the authors realized according to the latter solution, where two transducers were mounted below the keybed, one mid-way the keyboard length and another approximately at one quarter way the same length—see also Figure 4 (right).

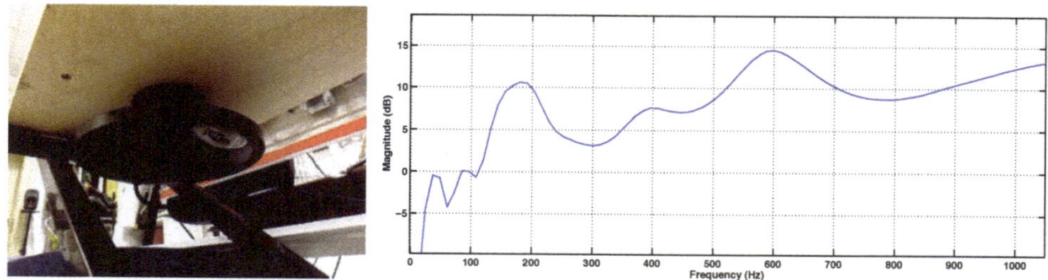

Figure 6. Left: tactile transducer mounted below the haptic digital piano keybed. **Right**: average spectral flattening curve. Pictures from [20].

By playing an impulse simultaneously on both of them, advanced BiVib users should accurately measure on each key the corresponding impulse response, and then (with Aurora or similar tools) design an equalizer which on average flattens the "coloured" response affecting the vibrations due to their path from the audio interface to the keyboard. Since such measures depend on the user's musical keyboard, they could not be included in BiVib as the binaural room transfer functions were instead.

However, in the aforementioned digital piano customization it was experimentally observed by the authors that the tactile transducers shown in Figure 6 (left) were mostly responsible for the inclusion of distinct spectral peaks affecting all the key responses. Hence, the average equalization curve shown in Figure 6 (right) was estimated, and proved to be effective for flattening the peaks measured on an evenly-spaced subset of the keys. A similar curve can be approximated by BiVib users by designing an equalizer that flattens the frequency response of the tactile transducers equipping the setup. This response is normally reported in a good quality transducer's data sheet.

4. Experiments, Applications and Future Work

BiVib has been originally created to support multisensory experiments in which precise control had to be maintained over the simultaneous auditory and vibrotactile stimuli reaching a performing pianist, particularly when judging the perceived quality of an instrument. In a recent paper, the authors were able to confirm subjective vibrotactile frequency thresholds of active touch [34] by conducting tests with pianists who were asked to detect vibrations at the piano keyboard [12]. A comparison between such thresholds and the spectrum of a lower fortissimo A0 tone is shown in Figure 7. This demonstrates a previously unreported results: during active playing, pianists are able to integrate tactile sensation that would be imperceptible in passive touch conditions [38].

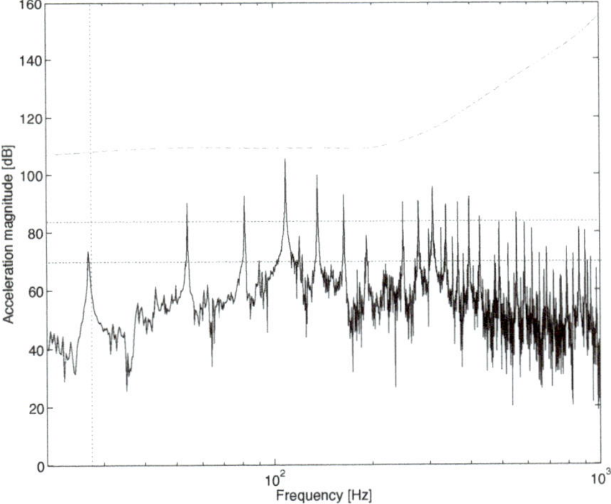

Figure 7. Magnitude spectrum of the vibration signal at the A0 key, recorded on the upright Disklavier playing at MIDI velocity 111. The dash-dotted curve depicts the reference vibrotactile threshold for passive touch [38], while the two horizontal dashed lines represent the minimum and maximum thresholds recently measured under active touch conditions [34]. Picture adapted from [12].

A precise manipulation of the intensity relations between piano sound and vibrations may be used to investigate the existence of cross-modal effects occurring during piano playing. Such effects have been discovered as part of a more general multisensory integration mechanism [39] that under certain conditions can increase the perceived intensity of auditory signals [40], or conversely enhance touch perception [41]. Aiming to understand whether piano keyboard vibrations impact the perceived quality of the instrument and, as a secondary effect, the quality of a performance, the authors have first observed significant differences in the perceived quality of vibrating vs. silently playing Disklavier pianos. This observation marks a point in favor of the former, especially since that pianists involved in the experiment reported to be unaware of the existence of vibratory feedback [12]. In other words, while most pianists preferred the vibrating instrument, they did not consciously realize that their decision was caused by the vibrations produced (or not produced) by the instrument during the performance. Based on this result, a further experiment was designed for the haptic digital piano using diverse types of tactile feedback, synthesized by manipulating the BiVib samples. The test aimed at investigating possible consequences of the vibrotactile feedback on the pianist's playing experience (qualitative effect) and on the performance in terms of timing and dynamics accuracy (quantitative effect). Cross-modal effects resulting from varying the tactile feedback of the keyboard were observed, still these preliminary results are far from giving a systematic view about the impact of the different sensory channels on the pianist's playing experience, and especially on the accuracy of execution [20].

Another potential use of BiVib is in the investigation of binaural spatial cues for the acoustic piano. Using the recommendations given in the previous section, single tones of the upright piano can in fact be accurately cleared of the room echoes, and then be reproduced in the position of the ear entrance. The existence of localization cues in piano sounds has not been completely understood yet. Even in pianos where these cues are reported to be audible by listeners, their exact acoustic origin is still an open question [42]. Moreover, visual cues of self-moving keys (a condition possible on Disklavier pianos) producing the corresponding tones, as well as somatosensory cues occurring during active piano playing, may have an influence on localization judgments [30,32].

One further direction which may take advantage of BiVib deals with research in cognitive neuroscience: recently, pianists and their instrument served as key subjects for understanding diverse aspects of brain and motor development [43,44]. In this context, the contribution of the auditory, visual and tactile sensory modalities to this development have not been ascertained yet. Such knowledge could be of help not only to capture more general aspects of the development of human senses, but also to guide a perceptually and cognitively informed design of novel keyboard interfaces.

In summary, future research that can be conducted in the laboratory using BiVib includes tests aiming at conclusively understanding whether (i) vibrations affect the performance on the keyboard, and whether (ii) auditory lateralization is able to guide piano tone localization. If so, such multisensory cues may substantially contribute to the sense of engagement and, hence, improve the quality of the performance and make the learning curve of a keyboard interface more acceptable.

Outside the laboratory, the library can reward musicians who simply wish to use its sounds. On the one hand, the grand piano recordings contain echoes that bring distinct cues of the room where they have been recorded. In this sense they are ready for use, although labeled by a precise acoustic footprint. On the other hand, the upright piano recordings are much more anechoic and, hence, far less difficult to spatialize than piano recordings taken outside an acoustically controlled room [45]. Consequently, they can be easily imported and conveniently personalized by users through artificial reverberation.

5. Conclusions

BiVib provides a unique set of multimodal piano data recorded using high-quality equipment in controlled conditions through reproducible computer-controlled procedures. Since its original release, the library has been enriched with binaural responses of the room where the upright piano was recorded. We hope that a successful use of the BiVib dataset, in conjunction with this documentation and through publicly available projects for the Kontakt software sampler, will facilitate further research in piano acoustics, performance, and new musical interface design also for educational purposes.

Author Contributions: Conceptualization, F.A., F.F. and S.P.; methodology, F.A., F.F. and S.P.; software, S.P.; resources, F.A., F.F. and S.P.; data curation, S.P.; writing—original draft preparation, S.P.; writing—review and editing, F.A., F.F. and S.P.

Funding: This research was partially funded by Swiss National Science Foundation grants number 150107 and 178972.

Acknowledgments: This research received support by project AHMI (Audio-haptic modalities in musical interfaces, 2014–2016), and HAPTEEV (Haptic technology and evaluation for digital musical interfaces 2018–2022), both funded by the Swiss National Science Foundation. Support came also from the PRID project SMARTLAND, funded by the University of Udine. The Disklavier grand model DC3 M4 located in Padova was made available by courtesy of the Sound and Music Processing Lab (SaMPL), a project of the Conservatory of Padova funded by Cariparo Foundation (thanks in particular to Nicola Bernardini and Giorgio Klauer). Gianni Chiaradia donated the wooden panel used to build the haptic digital piano. The authors would like to thank several students and collaborators who contributed to the development of this work along the years, in chronological order: Francesco Zanini, Valerio Zanini, Andrea Ghirotto, Devid Bianco, Lorenzo Malavolta, Debora Scappin, Mattia Bernardi, Francesca Minchio, Martin Fröhlich.

Conflicts of Interest: The author declares no conflict of interest.

References

1. Turner, E.O. Touch and Tone-Quality: The Pianist's Illusion. *Music. Times* **1939**, *80*, 173–176. [CrossRef]
2. Palmer, C.F.; Jones, R.K.; Hennessy, B.L.; Unze, M.G.; Pick, A.D. How is a trumpet known? The "basic object level" concept and perception of musical instruments. *Am. J. Psychol.* **1989**, *102*, 17–37. [CrossRef] [PubMed]
3. Galembo, A.; Askenfelt, A. Quality assessment of musical instruments—Effects of multimodality. In Proceedings of the 5th Triennial Conference of the European Society for the Cognitive Sciences of Music (ESCOM5), Hannover, Germany, 8–13 September 2003.
4. Hodges, D.A.; Hairston, W.D.; Burdette, J.H. Aspects of Multisensory Perception: The Integration of Visual and Auditory Information in Musical Experiences. *Ann. N. Y. Acad. Sci.* **2005**, *1060*, 175–185. [CrossRef] [PubMed]
5. Campbell, D.M. Evaluating musical instruments. *Phys. Today* **2014**, *67*, 35–40. [CrossRef]
6. Marshall, K.; Genter, B. The musician and the vibrational behavior of a violin. *J. Catgut Acoust. Soc.* **1986**, *45*, 28–33.
7. Suzuki, H. Vibration and sound radiation of a piano soundboard. *J. Acoust. Soc. Am.* **1986**, *80*, 1573–1582. [CrossRef]
8. Askenfelt, A.; Jansson, E.V. On Vibration Sensation and Finger Touch in Stringed Instrument Playing. *Music Percept.* **1992**, *9*, 311–349. [CrossRef]
9. Keane, M.; Dodd, G. Subjective Assessment of Upright Piano Key Vibrations. *Acta Acust. United Acust.* **2011**, *97*, 708–713. [CrossRef]
10. Saitis, C. Evaluating Violin Quality: Player Reliability and Verbalization. Ph.D. Thesis, Department of Music Research, McGill University, Montreal, QC, Canada, 2013.
11. Wollman, I.; Fritz, C.; Poitevineau, J. Influence of vibrotactile feedback on some perceptual features of violins. *J. Acoust. Soc. Am.* **2014**, *136*, 910–921. [CrossRef] [PubMed]
12. Fontana, F.; Papetti, S.; Järveläinen, H.; Avanzini, F. Detection of keyboard vibrations and effects on perceived piano quality. *J. Acoust. Soc. Am.* **2017**, *142*, 2953–2967. [CrossRef] [PubMed]
13. O'Modhrain, S.; Gillespie, B.R. Once More, with Feeling: The Dynamics of Performer-Instrument Interaction. In *Musical Haptics*; Papetti, S., Saitis, C., Eds.; Springer: Berlin, Germany, 2018.
14. Papetti, S.; Saitis, C. (Eds.) *Musical Haptics*; Springer: Berlin, Germany, 2018.
15. O'Modhrain, S.; Chafe, C. Incorporating Haptic Feedback into Interfaces for Music Applications. In Proceedings of the World Automation Conference ISORA, Maui, HI, USA, 11–16 June 2000.
16. Marshall, M.T.; Wanderley, M.M. Vibrotactile feedback in digital musical instruments. In Proceedings of the 2006 Conference on New Interfaces for Musical Expression, Paris, France, 4–8 June 2006; pp. 226–229.
17. Overholt, D.; Berdahl, E.; Hamilton, R. Advancements in Actuated Musical Instruments. *Organ. Sound* **2011**, *16*, 154–165. [CrossRef]
18. Birnbaum, D.M.; Wanderley, M.M. A systematic approach to musical vibrotactile feedback. In Proceedings of the International Computer Music Conference, Copenhagen, Denmark, 27–31 August 2007.
19. Michailidis, T.; Bullock, J. Improving performers' musicality through live interaction with haptic feedback: A case study. In Proceedings of the Sound and Music Computing (SMC), Padova, Italy, 6–9 July 2011.
20. Fontana, F.; Avanzini, F.; Järveläinen, H.; Papetti, S.; Klauer, G.; Malavolta, L. Rendering and Subjective Evaluation of Real vs. Synthetic Vibrotactile Cues on a Digital Piano Keyboard. In Proceedings of the Sound and Music Computing Conference, Maynooth, Ireland, 26 July–1 August 2015; pp. 161–167.
21. Young, G.W.; Murphy, D.; Weeter, J. A Qualitative Analysis of Haptic Feedback in Music Focused Exercises. In Proceedings of the New Interfaces for Musical Expression NIME'17, Copenhagen, Denmark, 15–18 May 2017; pp. 204–209.

22. Maestre, E.; Papiotis, P.; Marchini, M.; Llimona, Q.; Mayor, O.; Pérez, A.; Wanderley, M.M. Enriched Multimodal Representations of Music Performances: Online Access and Visualization. *IEEE MultiMed.* **2017**, *24*, 24–34. [CrossRef]

23. Li, B.; Liu, X.; Dinesh, K.; Duan, Z.; Sharma, G. Creating a Multitrack Classical Music Performance Dataset for Multimodal Music Analysis: Challenges, Insights, and Applications. *IEEE Trans. Multimed.* **2019**, *21*, 522–535. [CrossRef]

24. Koelstra, S.; Muhl, C.; Soleymani, M.; Lee, J.S.; Yazdani, A.; Ebrahimi, T.; Pun, T.; Nijholt, A.; Patras, I. Deap: A database for emotion analysis; using physiological signals. *IEEE Trans. Affect. Comput.* **2012**, *3*, 18–31. [CrossRef]

25. Von Coler, H. TU-Note Violin Sample Library—A Database of Violin Sounds with Segmentation Ground Truth. In Proceedings of the 21st International Conference on Digital Audio Effects (DAFx-18), Aveiro, Portugal, 4–8 September 2018; pp. 312–317.

26. Mayor, O.; Llop, J.; Maestre Gómez, E. RepoVizz: A multi-modal on-line database and browsing tool for music performance research. In Proceedings of the 12th International Society for Music Information Retrieval Conference (ISMIR 2011), Miami, FL, USA, 24–28 October 2011.

27. Moog, R.A.; Rhea, T.L. Evolution of the Keyboard Interface: The Bösendorfer 290 SE Recording Piano and the Moog Multiply-Touch-Sensitive Keyboards. *Comput. Music J.* **1990**, *14*, 52–60. [CrossRef]

28. McPherson, A.P.; Gierakowski, A.; Stark, A.M. The space between the notes: Adding expressive pitch control to the piano keyboard. In Proceedings of the SIGCHI Conference on Human Factors in Computing Systems, Paris, France, 27 April–2 May 2013; ACM: New York, NY, USA, 2013; pp. 2195–2204.

29. d'Alessandro, N.; Tilmanne, J.; Moreau, A.; Puleo, A. AirPiano: A Multi-Touch Keyboard with Hovering Control. In *Proceedings of the International Conference on New Interfaces for Musical Expression*; Berdahl, E., Allison, J., Eds.; Louisiana State University: Baton Rouge, LA, USA, 2015; pp. 255–258.

30. Fontana, F.; Scappin, D.; Avanzini, F.; Bernardi, M.; Bianco, D.; Klauer, G. Auditory, Visual And Somatosensory Localization Of Piano Tones: A Preliminary Study. In Proceedings of the Sound Music Computing Conference (SMC), Espoo, Finland, 5–8 July 2017; pp. 254–260.

31. Papetti, S.; Avanzini, F.; Fontana, F. BIVIB: A Multimodal Piano Sample Library of Binaural Sounds and Keyboard Vibrations. In Proceedings of the International Conference on Digital Audio Effects (DAFx-18), Aveiro, Portugal, 4–8 September 2018; pp. 237–243.

32. Fontana, F.; Avanzini, F.; Papetti, S. Evidence of lateralization cues in grand and upright piano sounds. In Proceedings of the 15th International Conference on Sound and Music Computing (SMC2018), Limassol, Cyprus, 4–7 July 2018; pp. 80–84.

33. Goebl, W.; Bresin, R. Measurement and reproduction accuracy of computer-controlled grand pianos. *J. Acoust. Soc. Am.* **2003**, *114*, 2273–2283. [CrossRef] [PubMed]

34. Papetti, S.; Järveläinen, H.; Giordano, B.L.; Schiesser, S.; Fröhlich, M. Vibrotactile sensitivity in active touch: Effect of pressing force. *IEEE Trans. Haptics* **2017**, *10*, 113–122. [CrossRef] [PubMed]

35. Farina, A. Simultaneous Measurement of Impulse Response and Distortion with a Swept-Sine Technique. In Proceedings of the 108th Convention of the Audio Engineering Society, Paris, France, 19–22 February 2000.

36. Kirkeby, O.; Nelson, P.A.; Hamada, H.; Orduña-Bustamante, F. Fast deconvolution of multichannel systems using regularization. *IEEE Trans. Speech Audio Process.* **1998**, *6*, 189–194. [CrossRef]

37. Papetti, S.; Schiesser, S.; Fröhlich, M. Multi-point vibrotactile feedback for an expressive musical interface. In *Proceedings of the International Conference on New Interfaces for Musical Expression*; Berdahl, E., Allison, J., Eds.; Louisiana State University: Baton Rouge, LA, USA, 2015; pp. 235–240.

38. Verrillo, R.T. Vibration sensation in humans. *Music Percept.* **1992**, *9*, 281–302. [CrossRef]

39. Kayser, C.; Petkov, C.I.; Augath, M.; Logothetis, N.K. Integration of Touch and Sound in Auditory Cortex. *Neuron* **2005**, *48*, 373–384. [CrossRef] [PubMed]

40. Gillmeister, H.; Eimer, M. Tactile enhancement of auditory detection and perceived loudness. *Brain Res.* **2007**, *1160*, 58–68. [CrossRef] [PubMed]

41. Ro, T.; Hsu, J.; Yasar, N.E.; Elmore, L.C.; Beauchamp, M.S. Sound enhances touch perception. *Exp. Brain Res.* **2009**, *195*, 135–143. [CrossRef] [PubMed]

42. Askenfelt, A. Observations on the transient components of the piano tone. *STL-QPSR* **1993**, *34*, 15–22. Available online: http://www.speech.kth.se/qpsr/show_by_author.php?author=Askenfelt%2C+A (accessed on 1 March 2019).

43. Kalra, S.; Jain, S.; Agarwal, A. Infrared Powered Vibro-Haptic Piano Training System for the Visually Impaired. *Int. J. Adv. Soft Comput. Appl.* **2017**, *9*, 49–68.
44. Alves-Pinto, A.; Ehrlich, S.; Cheng, G.; Turova, V.; Blumenstein, T.; Lampe, R. Effects of short-term piano training on measures of finger tapping, somatosensory perception and motor-related brain activity in patients with cerebral palsy. *Neuropsychiatr. Dis. Treat.* **2017**, *13*, 2705. [CrossRef] [PubMed]
45. Rocchesso, D. Spatial effects. In *Digital Audio Effects*; Zölzer, U., Ed.; John Wiley & Sons: Chirchester Sussex, UK, 2002; pp. 137–200.

Article

Analysis and Modeling of Timbre Perception Features in Musical Sounds

Wei Jiang [1,2,3], Jingyu Liu [1,2,3], Xiaoyi Zhang [1,2,3], Shuang Wang [1,2,3] and Yujian Jiang [1,2,3,*]

[1] Key Laboratory of Acoustic Visual Technology and Intelligent Control System, Communication University of China, Ministry of Culture and Tourism, Beijing 100024, China; jw@cuc.edu.cn (W.J.); drumking@126.com (J.L.); zx_y@cuc.edu.cn (X.Z.); wsluckystar@126.com (S.W.)

[2] Beijing Key Laboratory of Modern Entertainment Technology, Communication University of China, Beijing 100024, China

[3] School of Information and Communication Engineering, Communication University of China, Beijing 100024, China

* Correspondence: yjjiang@cuc.edu.cn

Received: 25 December 2019; Accepted: 20 January 2020; Published: 22 January 2020

Abstract: A novel technique is proposed for the analysis and modeling of timbre perception features, including a new terminology system for evaluating timbre in musical instruments. This database consists of 16 expert and novice evaluation terms, including five pairs with opposite polarity. In addition, a material library containing 72 samples (including 37 Chinese orchestral instruments, 11 Chinese minority instruments, and 24 Western orchestral instruments) and a 54-sample objective acoustic parameter set were developed as part of the study. The method of successive categories was applied to each term for subjective assessment. A mathematical model of timbre perception features (i.e., bright or dark, raspy or mellow, sharp or vigorous, coarse or pure, and hoarse or consonant) was then developed for the first time using linear regression, support vector regression, a neural network, and random forest algorithms. Experimental results showed the proposed model accurately predicted these attributes. Finally, an improved technique for 3D timbre space construction is proposed. Auditory perception attributes for this 3D timbre space were determined by analyzing the correlation between each spatial dimension and the 16 timbre evaluation terms.

Keywords: feature extraction; timbre modeling; auditory perception; timbre space

1. Introduction

The subjective perception of sound originates from three auditory attributes: loudness, pitch, and timbre [1]. In recent years, researchers have established relatively mature evaluation models for loudness and pitch [2,3], but a quantitative calculation and assessment of timbre is far more complicated. Studies have shown that timbre is a critical acoustic cue for conveying musical emotion. It also provides an important basis for human recognition and classification of music, voice, and ambient sounds [4]. Therefore, the quantitative analysis of timbre and the establishment of a parameterized model are of significant interest in the fields of audio-visual information processing, music retrieval, and emotion recognition. The subjective nature of timbre complicates the evaluation process, which typically relies on subjective evaluations, signal processing, and statistical analysis. The American National Standards Institute (ANSI) defines timbre as an attribute of auditory sensation in terms of which a listener can judge that two sounds similarly presented and having the same loudness and pitch are dissimilar [5], making it an important factor for distinguishing musical tones [6].

Timbre evaluation terms (i.e., timbre adjectives) are an important metric for describing timbre perception features. As such, a comprehensive and representative terminology system is critical for ensuring the reliability of experimental auditory perception data. Conventionally, timbre evaluation research has focused on the fields of music and language sound quality, traffic road noise control, automobile or aircraft engine noise evaluation, audio equipment sound quality design, and soundscape evaluation. Among these, research in English-speaking countries is relatively mature, as shown in Table 1. However, differences in nationality, cultural background, customs, language, and environment inevitably affect the cognition of timbre evaluation terms [7–11]. In addition, Chinese instruments differ significantly from Western instruments in terms of their structure, production material, and sound production mechanisms. The timbre of Chinese instruments is also more diverse than that of Western instruments and existing English timbre evaluation terms may not be sufficient for describing these nuances. As such, the construction of musical timbre evaluation terms is of great significance to the study of Chinese instruments.

Table 1. Previous studies on timbre evaluation terms.

Author	Year	Objects of Evaluation	Evaluation Terms
Solomon [12]	1958	20 different passive sonar sound	50 pairs
von Bismarck [13]	1974	35 voiced and unvoiced speech sounds, musical sounds	30 pairs
Pratt and Doak [14]	1976	Orchestral instrument (including string, woodwind, and brass)	19
Namba et al. [15]	1991	4 performances of the Promenades in "Pictures at an Exhibition"	60
Ethington and Punch [16]	1994	Sound generated by an electronic synthesizer	124
Faure et al. [17]	1996	12 synthetic Western traditional instrument sounds	23
Iwamiya and Zhan [9]	1997	24 music excerpts from CDs on the market	18 pairs
Howard and Tyrrell [18]	1997	Western orchestral instruments, tuning fork, organ, and softly sung sounds.	21
Shibuya et al. [19]	1999	"A" major scale playing on the violin (including 3 bow force, 3 bow speed, and 3 sounding point)	20
Kuwano et al. [20]	2000	48 systematically controlled synthetic auditory warning sounds	16 pairs
Disley and Howard [21]	2003	4 recordings of different organs	7
Moravec and Štepánek [22]	2003	Orchestra instrument (including bow, wind, and keyboard)	30
Collier [23]	2004	170 sonar sounds (including 23 different generating source types, 9 man-made, and 14 biological)	148
Martens and Marui [24]	2005	9 distorted guitar sound (including three nominal distortion types)	11 pairs
Disley et al. [25]	2006	12 instrument samples from the McGill university master samples (MUMS) library (including woodwind, brass, string, and percussion)	15
Stepánek [26]	2006	Violin sounds of tones B3, #F4, C5, G5, and D6 played using the same technique	25
Katz and Katz [27]	2007	Music recording work	27

Table 1. *Cont.*

Author	Year	Objects of Evaluation	Evaluation Terms
Howard et al. [28]	2007	12 acoustic instrument samples from the MUMS library, 3 from each of the 4 categories (including string, brass, woodwind, and percussion).	15
Barbot et al. [29]	2008	14 aircraft sounds (including departure and arrival)	90
Pedersen [30]	2008	Stimuli may be anything that evokes a response; such stimuli may stimulate one or many of the senses (e.g., hearing, vision, touch, olfaction, or taste)	631
Alluri and Toiviainen [31]	2010	One hundred musical excerpts (each with a duration of 1.5 s) of Indian popular music, including a wide range of genres such as pop, rock, disco, and electronic, containing various instrument combinations.	36 pairs
Fritz et al. [32]	2012	Violin sound	61
Altinsoy and Jekosch [33]	2012	Sounds of 24 cars in 8 driving conditions from different brands with different motorization to the participants	36
Elliott et al. [34]	2013	42 recordings representing the variety of instruments and include muted and vibrato versions where possible (included sustained tones at E-flat in octave 4)	16 pairs
Zacharakis et al. [35]	2014	23 sounds drawn from commonly used acoustic instruments, electric instruments, and synthesizers, with fundamental frequencies varying across three octaves	30
Skovenborg [36]	2016	70 recordings or mixes ranging from project-studio demos to commercial pre-masters, plus some live recordings, all from rhythmic music genres, such as pop and rock	30
Wallmark [37]	2019	Orchestral instruments (including woodwind, brass, string, and percussion)	50

Timbre contains complex information concerning the source of a sound. Humans can perform a series of tasks to recognize objects by listening to these sounds [38]. As such, the quantitative analysis and description of timbre perception characteristics has broad implications in military and civil fields, such as instrument recognition [39], music emotion recognition [40], singing quality evaluation [41], active sonar echo detection [42], and underwater target recognition [43]. Developing a mathematical model of timbre perception features is vital to achieving a quantitative description of timbre. Two primary methods have conventionally been used to quantify timbre perception features. The first is the concept of psychoacoustic parameters [6]. That is, by analyzing the auditory characteristics of the human ear, a mathematical model can be established to represent subjective feelings, such as sharpness, roughness, and fluctuation strength [44]. Since most of the experimental stimulus signals in these experiments were noise, the calculated value for the musical signal differed from the subjective feeling, which is both limited and one-sided. Another technique combines subjective evaluation experiments with statistical analysis. In other words, the experiment is designed according to differences in perceived features from sound signals, from which objective parameters can be extracted. The correlation between objective parameters and perceived features is established through statistical analysis or machine learning, which is then used to develop a mathematical model of the perceived features. This approach has been widely used in the fields of timbre modeling [45,46], music information retrieval [47], instrument classification [48], instrument consonance evaluation [49], interior car sound evaluation [50], and underwater target recognition [42]. However, the experimental materials in these studies were Western instruments or noise. Chinese instruments are unique in their mechanisms of sound production and playing techniques, producing a rich timbre variety. As such, it is necessary to use Chinese instruments as a stimulus to establish a more complete timbre perception model.

Timbre is an auditory attribute with multiple dimensions, which can be represented by a continuous timbre space. This structure is of great importance to the quantitative analysis and classification of sound properties. The semantic differential method was used in early timbre space research [12,13]. Recently, multidimensional scaling (MDS) based on dissimilarity has been used to construct these spaces. For example, Grey used 16 Western instrument sound samples to create a three-dimensional (3D) timbre space [51]. McAdams et al. studied the common dimensions of timbre spaces with synthetic sounds used as experimental materials, establishing a relationship between the dimensions of a space and the corresponding acoustic parameters [52]. Martens et al. used guitar timbre to study the differences in timbre spaces constructed under different language backgrounds [53,54]. Zacharakis and Pastiadis conducted a subjective evaluation and analysis using 16 Western musical instruments, proposing a luminance–texture–mass (LTM) model for semantic evaluation. In this process, six semantic scales were analyzed using principal component analysis (PCA) and multidimensional scaling (MDS) to produce two different timbre spaces [55]. Simurra and Queiroz used a set of 33 orchestral music excerpts that were subjectively rated using quantitative scales based on 13 pairs of opposing verbal attributes. Factor analysis was included to identify major perceptual categories associated with tactile and visual properties, such as mass, brightness, color, and scattering [56]. Multidimensional scaling requires the acquisition of a dissimilarity matrix between each sample. However, existing methods use a paired comparison technique for the subjective evaluation experiment. This approach not only involves a large experimental workload, it also imposes a higher professional requirement, making the evaluation scale difficult to control. This paper proposes a new indirect model for constructing timbre spaces based on the method of successive categories. In this system, the dissimilarity matrix is calculated based on experimental data from the method of successive categories. This reduces the workload and increases the stability and reliability of the data.

The remainder of this paper is organized as follows. Section 2 introduces the timbre library construction process and Section 3 develops the timbre evaluation terminology. Section 4 introduces the perception feature model, and the timbre space is constructed in Section 5. Section 6 concludes the paper. The research methodology for the study is presented in Figure 1.

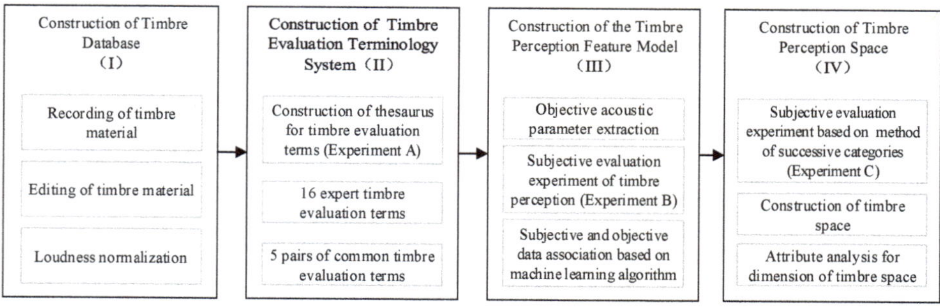

Figure 1. The proposed methodology.

2. Timbre Database Construction

2.1. Timbre Material Collection

A high-quality database of timbre materials was constructed by recording all materials required for the experiment in a full anechoic chamber, with a background noise level of −2 dBA. The equipment included a BK 4190 free-field microphone and a BK LAN-XI3560 AD converter. The performers were teachers and graduate students from the College of Music. Recordings consisted of musical scales and individual pieces of music. The Avid Pro Tools HD software was used to edit the audio material. The length of each clip was between 6–10 s, the sampling rate was 44,100 Hz, the quantization accuracy was 16 bits, and all audio was saved in the wav format. Previous studies on timbre used Western instruments as stimulus materials. However, the variety of timbre samples needed to be as rich as possible to increase the accuracy of timbre perception features. The timbre variety was enriched by using a collection of 72 different musical instruments, including 36 Chinese orchestral instruments, 12 Chinese minority instruments, and 24 Western orchestral instruments. The names and categories of the 72 instruments are listed in Appendix A. A timbre library containing 72 audio files was constructed from the data.

2.2. Loudness Normalization

In accordance with the definition of timbre, the influence of pitch and loudness are often excluded from timbre studies. However, previous research has shown that timbre and pitch are not independent in certain cases [57]. As such, timbre perception features presented in this paper include pitch as a factor. In order to eliminate the influence of loudness, a balance experiment was used to normalize the loudness of the timbre materials based on experimental results [58].

3. Construction of the Timbre Subjective Evaluation Term System

A timbre evaluation glossary including 32 evaluation terms was constructed and a subjective timbre evaluation experiment was conducted, based on a forced selection methodology (experiment A). Sixteen representative timbre evaluation terms were selected by combining the results of a clustering analysis. Finally, correlation analysis was used to calculate the correlation of these 16 evaluation terms. Six terms with a coefficient larger than 0.85 were removed. The remaining 10 terms were paired into five groups with opposite polarity (the absolute value of the correlation coefficient was greater than 0.81). These five pairs were used for timbre evaluation experiments based on the method of successive categories (experiment B), as well as the parametric modeling of timbre perception features.

3.1. Construction of the Thesaurus for Timbre Evaluation Terms

A thorough investigation of timbre evaluation terms was conducted under conditions of equivalent sound. A total of 329 terms were collected from the literature and a survey. Five people with a professional music background then deleted 155 of these terms (e.g., polysemy, ambiguous meaning, compound terms, etc.) that were, in their opinion, not suitable for a subjective experiment. A group of 21 music professionals listened to audio clips of the remaining 174 terms and judged whether they were suitable for describing the sound. The 32 most frequent evaluation terms were selected and a lexicon containing 32 timbre metrics was produced (Table 2). These terms completely describe all aspects of timbre dynamics, but they do include some redundant information, which needed to be assessed further using statistical analysis.

Table 2. A lexicon of 32 timbre evaluation terms in their original language (Chinese), with an accompanying English translation.

暗淡 (Dark)	饱满 (Plump)	纯净 (Pure)	粗糙 (Coarse)
丰满 (Full)	干瘪 (Raspy)	干涩 (Dry)	厚实 (Thick)
尖锐 (Sharp)	紧张 (Intense)	空洞 (Hollow)	明亮 (Bright)
生硬 (Rigid)	嘶哑 (Hoarse)	透亮 (Clear)	透明 (Transparent)
粗涩 (Rough)	单薄 (Thin)	低沉 (Deep)	丰厚 (Rich)
厚重 (Heavy)	浑厚 (Vigorous)	混浊 (Muddy)	尖利 (Shrill)
清脆 (Silvery)	柔和 (Mellow)	柔软 (Soft)	沙哑 (Raucous)
温暖 (Warm)	纤细 (Slim)	协和 (Consonant)	圆润 (Fruity)

3.2. Experiment A: A Subjective Evaluation Experiment Based on a Forced Selection Methodology

A subjective evaluation experiment was conducted in a standard listening room with a reverberation time of 0.3 s, which conforms to listening standards [59]. A total of 41 music professionals (21 males) participated in the experiment. Their ages ranged between 18 and 35 and they had no history of hearing loss. A forced selection methodology was employed in which audio clips from the material library were played in turn and subjects determined whether a given evaluation term was suitable for describing the audio clip. Clustering analysis and correlation analysis were then used to assess the experimental data (as discussed below), producing a music expert timbre evaluation term system (including 16 evaluation terms) and an ordinary timbre evaluation term system (including 5 pairs of evaluation terms with opposite polarity).

3.3. Data Analysis and Conclusion of Experiment A

A multidimensional scale was used to analyze the distance relationships for 32 evaluation terms in the two-dimensional space. The distance relationship between the 32 terms is shown in Figure 2. It is evident from Figure 2 that the distance between terms was small in some regions, indicating a high degree of correlation. In order to reduce the workload of subsequent timbre perception feature modeling, cluster analysis was used to further reduce the dimensionality of the evaluation terms. Figure 3 shows a cluster pedigree diagram calculated using a system clustering method. Using this diagram and the selection frequency obtained previously, the 32 terms were combined to produce 16 timbre evaluation terms (see Table 3). These 16 terms constituted the music expert timbre evaluation system used in the modeling of timbre spaces (experiment C).

Table 3. A musical expert timbre evaluation term system, including 16 timbre evaluation terms in their original language (Chinese) and the corresponding English translations.

暗淡 (Dark)	尖锐 (Sharp)	协和 (Consonant)	纯净 (Pure)
粗糙 (Coarse)	清脆 (Silvery)	纤细 (Slim)	单薄 (Thin)
丰满 (Full)	混浊 (Muddy)	柔和 (Mellow)	干瘪 (Raspy)
厚实 (Thick)	明亮 (Bright)	嘶哑 (Hoarse)	浑厚 (Vigorous)

A common timbre evaluation terminology system was then developed by calculating the Pearson correlation coefficient (PCC) for these 16 terms. The 6 terms with the highest correlation (PCC > 0.85) were excluded, resulting in a correlation matrix for the remaining 10 terms (Table 4). Terms with negative PCCs or large absolute values were selected from this matrix to form evaluation pairs with opposite meanings. These 10 terms were then combined to form five pairs (Table 5), constituting an ordinary timbre evaluation system. These pairs were used for the timbre evaluation experiment based on the method of successive categories (experiment B) and the parametric modeling of timbre perception features.

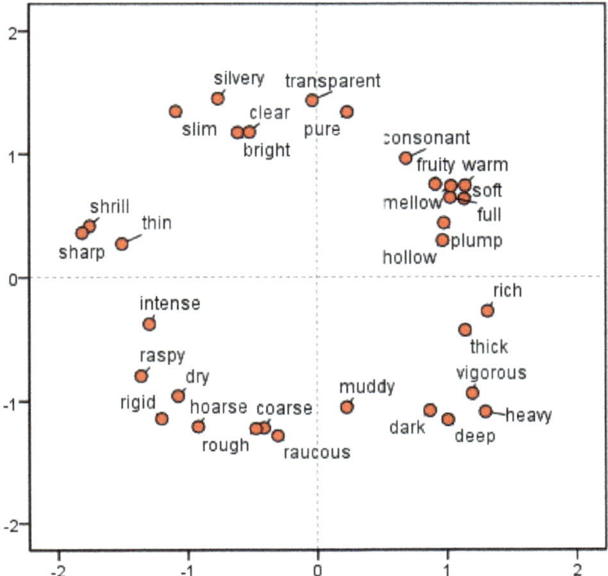

Figure 2. The distance relationship between the 32 evaluation terms.

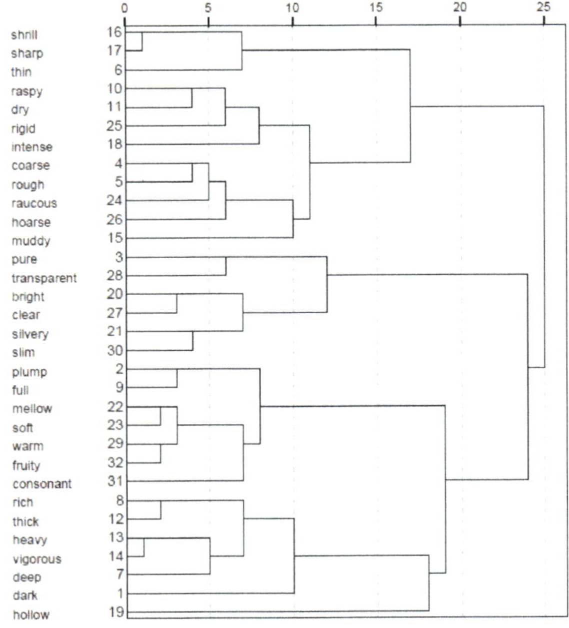

Figure 3. A cluster diagram of 32 timbre evaluation terms.

Table 4. A correlation matrix for 10 timbre evaluation terms.

	Bright	Dark	Sharp	Vigorous	Raspy	Coarse	Hoarse	Consonant	Mellow	Pure
Bright	1.00	−0.99	0.90	−0.93	0.24	−0.48	−0.31	0.13	−0.27	0.47
Dark	−0.99	1.00	−0.89	0.93	−0.20	0.49	0.33	−0.17	0.26	−0.48
Sharp	0.90	−0.89	1.00	−0.93	0.58	−0.14	0.06	−0.24	−0.57	0.17
Vigorous	−0.93	0.93	−0.93	1.00	−0.43	0.31	0.09	0.06	0.37	−0.28
Raspy	0.24	−0.20	0.58	−0.43	1.00	0.61	0.74	−0.83	−0.82	−0.51
Coarse	−0.48	0.49	−0.14	0.31	0.61	1.00	0.89	−0.82	−0.55	−0.92
Hoarse	−0.31	0.33	0.06	0.09	0.74	0.89	1.00	−0.86	−0.62	−0.83
Consonant	0.13	−0.17	−0.24	0.06	−0.83	−0.82	−0.86	1.00	0.79	0.75
Mellow	−0.27	0.26	−0.57	0.37	−0.82	−0.55	−0.62	0.79	1.00	0.51
Pure	0.47	−0.48	0.17	−0.28	−0.51	−0.92	−0.83	0.75	0.51	1.00

Table 5. An ordinary timbre evaluation term system including five pairs of evaluation terms in their original language (Chinese) and the associated English translations.

Name	Correlation Coefficient
明亮–暗淡 (Bright–Dark)	−0.99
干瘪–柔和 (Raspy–Mellow)	−0.82
尖锐–浑厚 (Sharp–Vigorous)	−0.93
粗糙–纯净 (Coarse–Pure)	−0.92
嘶哑–协和 (Hoarse–Consonant)	−0.86

4. Construction of a Timbre Perception Feature Model

Objective acoustic parameters were extracted from audio samples in 166 dimensions. The method of successive categories was then used to conduct a timbre perception evaluation experiment (experiment B), as well as reliability and validity analysis for the resulting data. Linear regression, support vector regression, a neural network, and a random forest algorithm were used to construct a timbre perception feature model. The accuracy of this model was then evaluated and it was used to predict timbre perception features for new audio materials.

4.1. Construction of the Objective Acoustic Parameter Set

Timbre is a multidimensional perception attribute that is closely related to the time-domain waveform and spectral structure of sound [60]. In order to establish a timbre perception feature model, an objective acoustic parameter set was constructed using 54 parameters extracted from the timbre database. Objective acoustic parameters refer to any values acquired using a mathematical model representing a normal sound signal in the time and frequency domains. These 54 parameters can be divided into 6 categories [61]:

(1) *Temporal shape features:* calculated from the waveform or the signal energy envelope (e.g., attack-time, temporal increase or decrease, and effective duration).

(2) *Temporal features:* auto-correlation coefficients with a zero-crossing rate.

(3) *Energy features:* referring to various energy content in the signal (i.e., global energy, harmonic energy, or noise energy).

(4) *Spectral shape features:* calculated from the short-time Fourier transform (STFT) of the signal (e.g., centroid, spread, skewness, kurtosis, slope, roll-off frequency, or Mel-frequency cepstral coefficients).

(5) *Harmonic features:* calculated using sinusoidal harmonic modeling of the signal (e.g., harmonic/ noise ratio, odd-to-even and tristimulus harmonic energy ratio, and harmonic deviation).

(6) *Perceptual features:* calculated using a model of human hearing (i.e., relative specific loudness, sharpness, and spread).

4.2. Calculation Method

The acoustic parameters were calculated as follows. The spectral centroid for the magnitude spectrum of the STFT [60] is given by:

$$C_t = \frac{\sum\limits_{n=1}^{N} M_t[n] \times n}{\sum\limits_{n=1}^{N} M_t[n]}, \tag{1}$$

where $M_t[n]$ is the magnitude of the Fourier transform at frame t and frequency n. This centroid is a measure of the spectral shape, where higher centroid values indicate "brighter" sounds. Spectral slope was calculated using a linear regression over spectral amplitude values. It should be noted that spectral slope is linearly dependent on the spectral centroid as follows [62]:

$$\text{slope}(t_m) = \frac{1}{\sum\limits_{k=1}^{K} a_k(t_m)} \times \frac{K \sum\limits_{k=1}^{K} f_k \cdot a_k(t_m) - \sum\limits_{k=1}^{K} f_k \cdot \sum\limits_{k=1}^{K} a_k(t_m)}{K \sum\limits_{k=1}^{K} f_k^2 - \left(\sum\limits_{k=1}^{K} f_k \right)^2}, \tag{2}$$

where $\text{slope}(t_m)$ is the spectral slope at time t_m, a_k is the spectral amplitude at k, and f_k is the frequency at k. Tristimulus values were introduced by Pollard and Jansson as a timbral equivalent to color attributes in vision. The tristimulus comprises three different energy ratios, providing a description of the first harmonics in a spectrum [63]:

$$T1(t_m) = \frac{a_1(t_m)}{\sum\limits_{h=1}^{H} a_h(t_m)},$$

$$T2(t_m) = \frac{a_2(t_m) + a_3(t_m) + a_4(t_m)}{\sum\limits_{h=1}^{H} a_h(t_m)}, \tag{3}$$

$$T3(t_m) = \frac{\sum\limits_{h=5}^{H} a_h(t_m)}{\sum\limits_{h=1}^{H} a_h(t_m)},$$

where H is the total number of partials and a_h is the amplitude of partial h.

Spectral flux is a time-varying descriptor calculated using STFT magnitudes. It represents the degree of variation in a spectrum over time, defined as unity minus the normalized correlation between successive a_k terms [64]:

$$\text{spectral flux} = 1 - \frac{\sum\limits_{k=1}^{K} a_k(t_{m-1}) a_k(t_m)}{\sqrt{\sum\limits_{k=1}^{K} a_k(t_{m-1})^2} \sqrt{\sum\limits_{k=1}^{K} a_k(t_m)^2}}. \tag{4}$$

Inharmonicity measures the departure of partial frequencies f_h from purely harmonic frequencies hf_0. It is calculated as a weighted sum of deviations from harmonicity for each individual partial [62]:

$$\text{inharmo}(t_m) = \frac{2}{f_0(t_m)} \frac{\sum\limits_{h=1}^{H} (f_h(t_m) - h f_0(t_m)) a_h^2(t_m)}{\sum\limits_{h=1}^{H} a_h^2(t_m)}, \tag{5}$$

where f_0 is the fundamental frequency and f_h is the frequency of partial h.

Spectral roll-off was proposed by Scheirer and Slaney [65]. It is defined as the frequency $f_c(t_m)$ below which 95% of the signal energy is contained:

$$\sum_{f=0}^{f_c(t_m)} a_f^2(t_m) = 0.95 \sum_{f=0}^{sr/2} a_f^2(t_m), \tag{6}$$

where $sr/2$ is the Nyquist frequency and a_f is the spectral amplitude at frequency f. In the case of harmonic sounds, it can be shown experimentally that spectral roll-off is related to the harmonic or noise cutoff frequency. The spectral roll-off also reveals an aspect of spectral shape as it is related to the brightness of a sound.

The odd-to-even harmonic energy ratio distinguishes sounds with a predominant energy at odd harmonics (such as the Guan) from other sounds with smoother spectral envelopes (such as the Suona). It is defined as:

$$\text{OER}(t_m) = \frac{\displaystyle\sum_{h=1}^{H/2} a_{2h-1}^2(t_m)}{\displaystyle\sum_{h=1}^{H/2} a_{2h}^2(t_m)}. \tag{7}$$

Twelve time-varying statistics were calculated for the 54 parameters, including the maximum, minimum, mean, variance, standard deviation, interquartile range, skewness coefficient, and kurtosis coefficient, producing an objective acoustic parameter set containing 166 dimensions (see Table 6). In this paper, Timbre Toolbox [62] and MIRtoolbox [66] were used for feature extraction. The corresponding acoustic parameters were extracted from materials in the timbre database and the acquired data were used to construct a timbre perception feature model.

4.3. Experiment B: A Timbre Evaluation Experiment Based on the Method of Successive Categories

A subjective evaluation experiment was conducted in a standard listening room with a reverberation time of 0.3 s, which conforms to listening standards [59]. A total of 34 subjects (16 males) with a professional music background participated in the experiment. Their ages ranged from 18 to 35 and they had no history of hearing loss. The experimental subjective evaluation process was conducted as follows. Material fragments were played, and the subjects judged the psychological scale of the piece for each timbre perception feature (evaluation term) in sequence, scoring it on a nine-level scale. All experimental materials were played prior to the formal experiment to familiarize subjects with the samples in advance. This was done to assist each subject in mastering the evaluation criteria and scoring scale, reducing the discretization of evaluation data for the same sample. Each piece was played twice with an interval of 5 s and a sample length of 6–10 s. Each evaluation term was tested for 10 min, with a 15-min break every half hour.

The validity and reliability of data from these 34 samples were analyzed to calculate a correlation coefficient between the scores for each subject. The Euclidean distance between the evaluation terms was calculated using cluster analysis to identify the two subjects with the largest difference in each group. Some subjects may not have had a sufficient understanding of the purpose of the experiment. Data from these subjects were excluded and not used for subsequent timbre perception feature modeling. The method of successive categories was used to conduct a statistical analysis of the experimental data [67]. The theoretical basis for this approach assumes the psychological scale to be a random variable, subject to a normal distribution. The boundary of each category was not a predetermined value, but a random variable identified from the experimental data. The Thurstone scale was then used to process the data and produce a psychological scale for all timbre materials and each perception feature for modeling purposes. Figure 4 shows the resulting scale for 72 musical instruments in 5 timbre evaluation dimensions. In each image, the dotted line represents the average value of each instrument in the corresponding dimension.

Table 6. Acoustic parameters.

Feature Name	Quantity	Feature Name	Quantity
Temporal Features		*Harmonic Spectral Shape*	
Log Attack Time	1	Harmonic Spectral Centroid	6
Temporal Increase	1	Harmonic Spectral Spread	6
Temporal Decrease	1	Harmonic Spectral Skewness	6
Temporal Centroid	1	Harmonic Spectral Kurtosis	6
Effective Duration	1	Harmonic Spectral Slope	6
Signal Auto-Correlation Function	12	Harmonic Spectral Decrease	1
Zero-Crossing Rate	1	Harmonic Spectral Roll-off	1
Energy Features		Harmonic Spectral Variation	3
Total Energy	1	*Perceptual Features*	
Total Energy Modulation	2	Loudness	1
Total Harmonic Energy	1	Relative Specific Loudness	24
Total Noise Energy	1	Sharpness	1
Spectral Features		Spread	1
Spectral Centroid	6	*Perceptual Spectral Envelope Shape*	
Spectral Spread	6	Perceptual Spectral Centroid	6
Spectral Skewness	6	Perceptual Spectral Spread	6
Spectral Kurtosis	6	Perceptual Spectral Skewness	6
Spectral Slope	6	Perceptual Spectral Kurtosis	6
Spectral Decrease	1	Perceptual Spectral Slope	6
Spectral Roll-off	1	Perceptual Spectral Decrease	1
Spectral Variation	3	Perceptual Spectral Roll-off	1
MFCC	12	Perceptual Spectral Variation	3
Delta MFCC	12	Odd-to-Even Band Ratio	3
Delta Delta MFCC	12	Band Spectral Deviation	3
Harmonic Features		Band Tristimulus	9
Fundamental Frequency	1	*Various Features*	
Fundamental Frequency Modulation	2	Spectral Flatness	4
Noisiness	1	Spectral Crest	4
Inharmonicity	1	*Total Number of Features*	166
Harmonic Spectral Deviation	3		
Odd-to-Even Harmonic Ratio	3		
Harmonic Tristimulus	9		

It is evident from Figure 4 that the distribution of timbre values for Chinese instruments differed significantly from Western instruments. For example, raspy/mellow and hoarse/consonant exhibited drastically different scales. This suggested the timbre database containing Chinese instruments had a richer variety of timbre types than a conventional Western instrument database. In addition, the distribution of timbre samples in the five timbre evaluation scale pairs was relatively balanced. This suggested the proposed evaluation terminology was representative of multiple timbre types and could better distinguish the attributes of different instruments. These factors could help to improve the accuracy of timbre perception feature models.

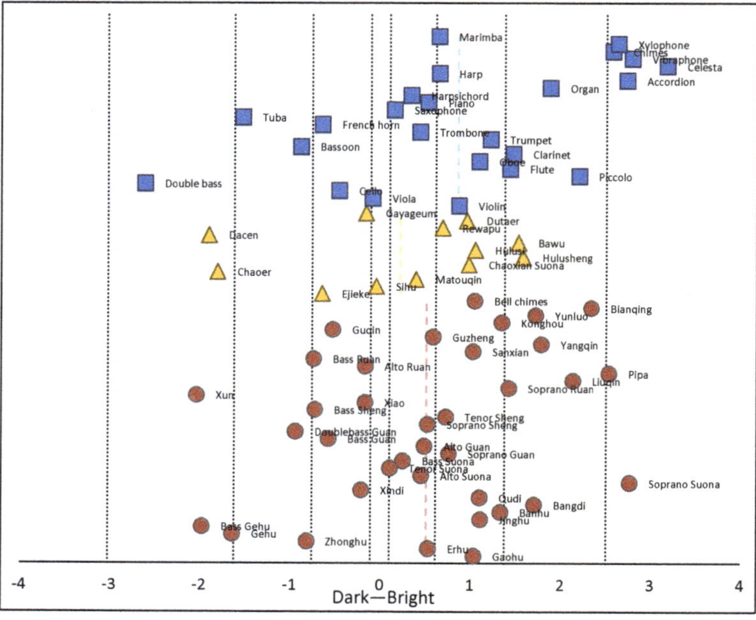

(**a**)

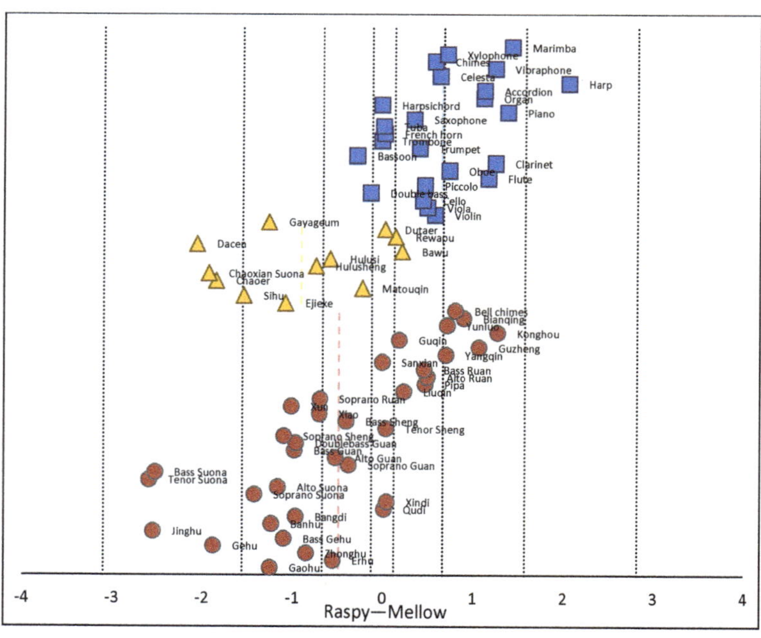

(**b**)

Figure 4. *Cont.*

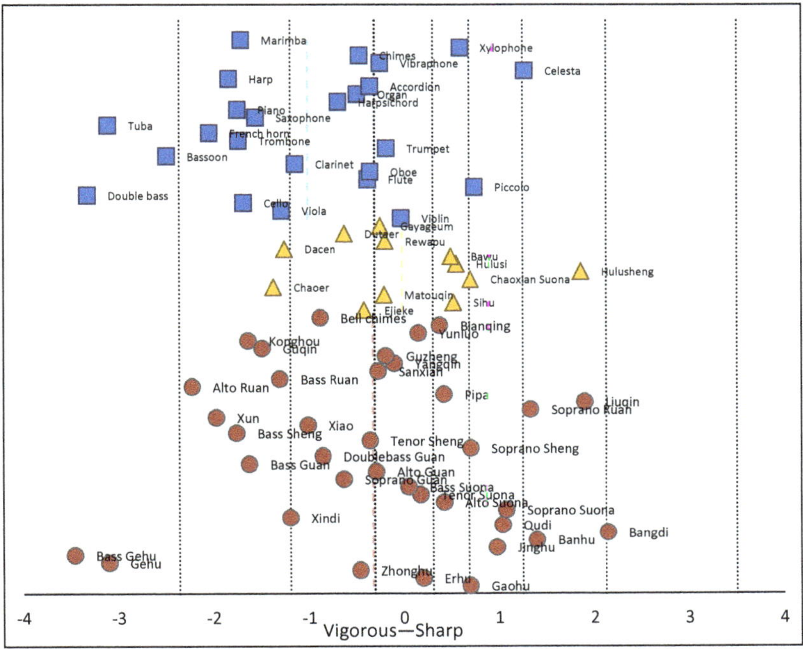

(**c**)

(**d**)

Figure 4. *Cont.*

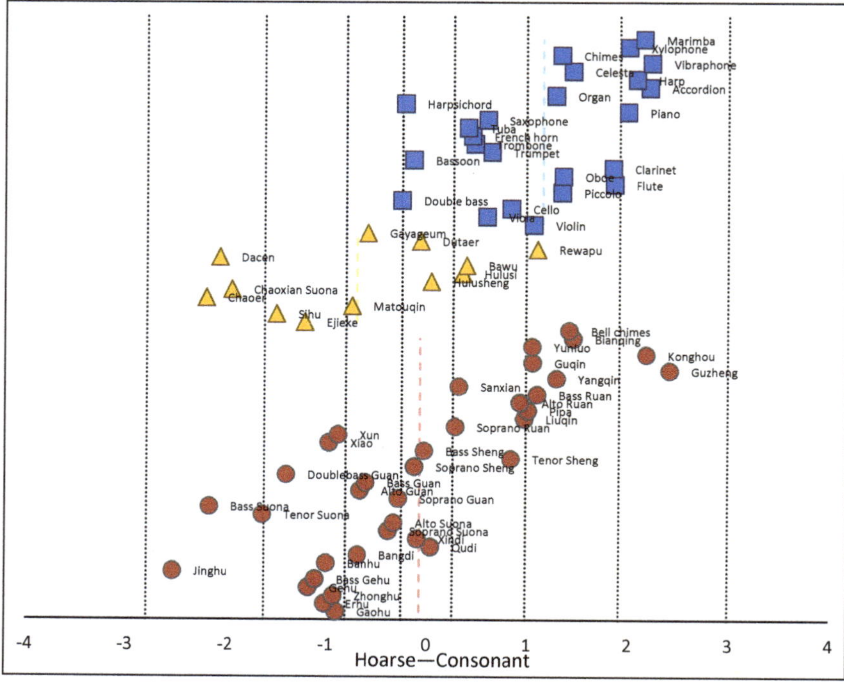

(e)

Figure 4. A psychological scale of 72 musical instruments, including (**a**) bright/dark, (**b**) raspy/mellow, (**c**) sharp/vigorous, (**d**) coarse/pure, and (**e**) hoarse/consonant. The blue squares represent Western orchestral instruments, the yellow triangles represent Chinese minority instruments, and the red circles represent Chinese orchestral instruments. The dotted blue line represents the mean value of the Western orchestral instruments, the dotted yellow line represents the mean value of the Chinese minority instruments, and the dotted red line represents the mean value of the Chinese orchestral instruments.

4.4. Construction of a Prediction Model

In this study, multiple linear regression, support vector regression, a neural network, and a random forest algorithm were used to correlate objective parameters and subjective evaluation experimental data to construct a mathematical model of timbre perception features. Stepwise techniques were used for variable entry and removal in the multiple linear regression algorithm [68], and radial basis functions were selected as kernels for support vector regression [69]. A multi-layer perceptron was adopted in the neural network, which included a hidden layer [70]. Random forest is a common ensemble model consisting of multiple CART-like trees, each of which grows on a bootstrap object acquired by sampling the original data cases with replacements [71].

Before modeling, feature selection was conducted for the target attribute to be predicted. This process consisted of three steps:

(1) *Screening*: removes unimportant or problematic predictors and cases.
(2) *Ranking*: sorts remaining predictors and assigns ranks; this step considers one predictor at a time to determine how well it predicts the target variable.
(3) *Selecting*: identifies the important subset of features to use in subsequent models.

During the modeling phase, 80% of the data were used for training and the remaining 20% were used for validation. The input to the model was a 166-dimensional objective parameter set and the output was the value of the five perception dimensions (bright/dark, raspy/mellow, sharp/vigorous, coarse/pure, and hoarse/consonant). Correlation coefficients were used to evaluate the accuracy of the model and represented the results of the correlation analysis between the model prediction data and subjective evaluation data, with higher coefficients representing a more accurate model.

The accuracy of prediction results for the four algorithms across the five perception dimensions are shown in Table 7. Figure 5 provides a histogram of the prediction accuracy in different dimensions. These experimental results suggested that the proposed technique provided valid predictions in each of the five dimensions. The algorithm exhibiting the best performance exceeded 0.9 for bright/dark, sharp/vigorous, coarse/pure, and hoarse/consonant sound types. The averaged results indicated that the neural network (0.915) and random forest (0.864) outperformed multiple linear regression (0.665) and support vector regression (0.670). The neural network was particularly accurate in its predictions of the five perception dimensions.

Table 7. A comparison of the accuracies achieved by four algorithms.

	Multiple Linear Regression	Support Vector Regression	Neural Network	Random Forest
Bright/Dark	0.706	0.696	0.913	0.856
Raspy/Mellow	0.573	0.571	0.858	0.813
Sharp/Vigorous	0.859	0.852	0.952	0.945
Coarse/Pure	0.481	0.518	0.928	0.827
Hoarse/Consonant	0.705	0.711	0.922	0.877
Average	0.665	0.670	0.915	0.864

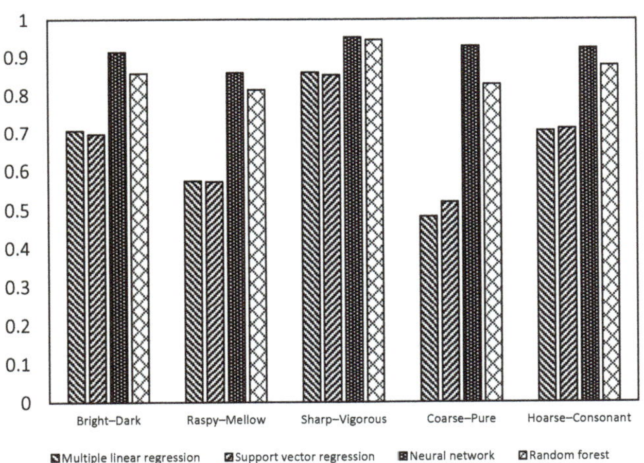

Figure 5. A prediction accuracy histogram for the five perception attributes.

5. The Construction of Timbre Space

Multidimensional scaling (MDS) was used to construct a 3D timbre perception space to represent the distribution of 37 Chinese instruments more intuitively. Unlike many common analysis methods, MDS is heuristic and does not require assumptions about spatial dimensionality [72]. It also offers the advantages of visualization and helps to identify potential factors affecting the similarity between terms. The construction of a timbre space includes three steps:

(1) *Subjective evaluation experiment based on sample dissimilarity:* where a dissimilarity matrix between samples was obtained using a subjective evaluation experiment. Existing research has conventionally paired up samples in the material database to score the dissimilarity. The process was simplified in this study, which reduced the workload.

(2) *Dimension reduction of distance matrix based on MDS:* where the MDS algorithm was used to calculate the dissimilarity matrix such that sample distances in high-dimensional spaces can be represented in low-dimensional spaces (usually two or three dimensions).

(3) *Attribute interpretation of each dimension of timbre space:* where the correlation between each dimension and the timbre perception features was analyzed using a statistical method. Interpretable attributes for each dimension were then acquired from this space.

The performance of multidimensional scaling algorithms depends on the sample dissimilarity matrix. In previous studies [51,52], this matrix was acquired using a subjective evaluation experiment that compared and scored the dissimilarity of any two samples. A total of $n^2/2$ experiments must be conducted for n samples. This quadratic relationship significantly increases the computational complexity and runtime, which makes quantifying the dissimilarity more difficult. This paper presents an improved methodology in which a set of evaluation indicators were selected (as complete as possible) and all samples were successively scored with each indicator. These results constituted the feature vector for the sample and the distance to each vector was calculated to obtain the dissimilarity of all samples. The 16 timbre evaluation terms shown in Table 3 were used to assess the attributes of each dimension during the analysis phase.

The method of successive categories was then used to conduct a subjective evaluation experiment on timbre materials for 37 Chinese instruments (experiment C). Grade 9 was performed on 16 perception dimensions in Table 3 and the reliability and validity of the experimental data were analyzed. The Euclidean distance of the feature vectors was calculated, producing a dissimilarity matrix for 37 Chinese instruments. The MDS algorithm was used to process the timbre dissimilarity matrix and construct a 3D timbre perception space.

5.1. Experiment C: Subjective Evaluation Experiment Based on Sample Dissimilarity

Three factors were considered during sample selection to prepare the sound data needed in the subjective evaluation experiment [73]:

(1) *The appropriate number of samples:* The number of samples must be sufficiently large to ensure the accuracy of the MDS algorithm and impose sufficient constraints on the model. In practice, however, it is difficult to establish precise rules for determining these data. However, empirical solutions do exist. In most MDS-based timbre studies, at least 10 sound samples are required for two-dimensional spaces and at least 15 sound samples are needed for three-dimensional spaces [51,74,75]. In this paper, 37 kinds of Chinese instruments were used as experimental materials, which ensured that sufficient constraints were provided to the MDS model.

(2) *The range of timbre variation:* The range of timbre varies depending on the subject of the study, with larger instrumental variety (i.e., orchestral music) providing better data [34]. Models constructed in this way can be applied more broadly to new timbre samples. In this study, 37 kinds of Chinese instruments were selected. As can be seen from Figure 4, compared with Western instruments, Chinese instruments had a wider distribution range in terms of their timbre evaluation scale. As such, the Chinese instrument samples selected in this paper ensured a diverse range of timbre samples.

(3) *The uniformity of timbre sample distributions:* The distribution of sound samples in each timbre perception attribute should be as uniform as possible. Timbre spaces are continuous perceptual spaces and a uniform distribution sample set is beneficial to the construction of continuous timbre spaces. Non-uniform sample distributions can degrade solutions to the MDS equations, preventing the structures between classes from being fully displayed [76]. As seen in Figure 4, the samples selected in this study covered a broad range of timbre attributes and they were distributed at varying psychological scales, providing a uniform distribution.

Subjective evaluation of the experimental environment and the subjects was conducted as in experiment B. The process was as follows: while playing each experimental sample, the subjects judged the psychological scale of the sample on 16 timbre perception features (timbre evaluation terms) in turn, scoring each on a 9-point scale.

5.2. The Construction of the 3D Timbre Space Using MDS

The reliability and validity processing method applied to the experimental data was the same as in experiment B. The processed data were averaged and the mean score for all subjects on each evaluation term was calculated for each sample. These data were then used to calculate the timbre dissimilarity, expressed in the form of a distance matrix. The MDS algorithm was adopted in this paper [77], which considers individual differences between subjects and assigns a corresponding weight to each score. This approach considers terms in every dimension and more fully utilizes the experimental data. Multidimensional scaling is based on dissimilarity analysis for two samples in a timbre attribute space, which can be expressed using a distance matrix as follows:

$$d^i_{jk} = \sqrt{\sum_{r=1}^{R} w_{ir} \cdot (x_{jr} - x_{kr})^2},$$

(8)

where d^i_{jk} represents the dissimilarity evaluation score for subject i assessing sounds j and k, w_{ir} represents the weight of subject i in the rth dimension, and x_{kr} represents the coordinates of sample k in the rth dimension.

Equation (8) was used to calculate the distance for 37 timbre feature vectors and the dissimilarity distance matrix for 37 samples. This matrix was used as input into the MDS algorithm. The number of timbre space dimensions was determined by referring to previous research results [51,52]. The timbre space dimension was determined in three dimensions using Kruskal's stress function [78]. The coordinates of each sound sample in 3D timbre space were acquired by using MDS to reduce the dimensionality of the dissimilarity distance matrix (Figure 6).

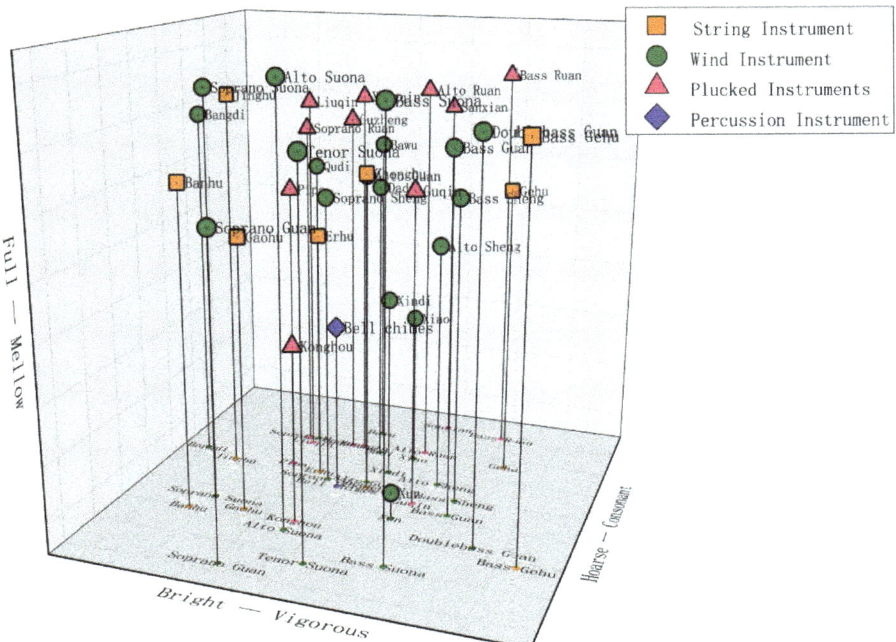

Figure 6. A 3D timbre space for 37 Chinese instruments.

5.3. Perception Attribute Analysis of the Timbre Space Dimension

The correlation between 16 timbre perception attributes was calculated to analyze the auditory attributes of each dimension in the timbre space. The coordinates of the samples were projected into three dimensions to obtain the spatial distribution of the data. Pearson correlation coefficients were calculated between each dimension and the 16 timbre perception attributes (Table 8). Further analysis suggested dimension 1 was positively correlated with the "bright" perception attribute and negatively correlated with "vigorous." As such, dimension 1 could be defined as "bright/vigorous." Dimension 2 was positively correlated with "hoarse" and negatively correlated with "consonant." However, the correlation of dimension 3 was not as obvious, as it was only slightly correlated with "full/mellow." Figure 6 suggests that different types of instruments were distributed at different positions in the timbre space, which could be used to categorize individual timbres.

Table 8. The results of correlation analysis in 3D timbre space.

Attribute	Dimension 1	Dimension 2	Dimension 3
纤细 (Slim)	0.97	−0.13	−0.11
明亮 (Bright)	0.97	−0.17	0.15
暗淡 (Dark)	−0.96	0.19	−0.14
尖锐 (Sharp)	0.95	0.23	0.14
浑厚 (Vigorous)	−0.99	−0.05	0.11
单薄 (Thin)	0.94	0.26	−0.10
厚实 (Thick)	−0.97	0.00	0.22
清脆 (Silvery)	0.96	−0.22	0.04
干瘪 (Raspy)	0.39	0.87	0.02
丰满 (Full)	−0.83	−0.38	0.33
粗糙 (Coarse)	−0.35	0.89	−0.06
纯净 (Pure)	0.34	−0.82	0.11
嘶哑 (Hoarse)	−0.15	0.93	−0.13
协和 (Consonant)	−0.02	−0.96	0.00
柔和 (Mellow)	−0.38	−0.80	−0.37
混浊 (Muddy)	−0.91	0.26	−0.16

6. Conclusions

This study presented a novel methodology for the analysis and modeling of timbre perception features in musical sounds. The primary contributions can be summarized as follows:

(1) A novel method was proposed for constructing two sets of timbre evaluation terminology systems in a Chinese context. Experimental results from a subjective evaluation showed that these terms could successfully distinguish timbre from different instruments.

(2) A timbre material library containing 72 musical instruments was constructed according to relevant standards. A subjective evaluation experiment was conducted using the method of successive categories. The psychological scales of the subjects were acquired using five pairs of perceptual dimensions. A mathematical model of timbre perception features was then developed using multiple linear regression, support vector regression, a neural network, and the random forest algorithm. Experimental results showed that this constructed model could predict perceptual features for new samples.

(3) An improved method for constructing 3D timbre space was proposed and demonstrated using the MDS algorithm applied to 37 Chinese instruments. Auditory perceptual attributes were determined by analyzing the correlation between the 3 dimensions of the timbre space and 16 perceptual attributes.

In future research, we will focus on the following three aspects of this study. First, supplemental sample materials will be acquired based on the existing timbre database. We will attempt to expand the variety and quantity of the data to improve the consistency and robustness of the model. Second, a subjective evaluation experiment, statistical analysis, and other techniques will be used to select timbre evaluation terms that accurately reflect the essential attributes of timbre to provide support for the construction of simple and effective timbre spaces. Third, the machine learning algorithm will be improved by including more subjective evaluation data. Additional correlation algorithms will also be tested to improve the accuracy of the model predictions. Finally, mathematical modeling will be implemented for each dimension in the timbre space. The distribution of other (i.e., Western) instruments will be compared to that of Chinese instruments to identify common patterns.

Author Contributions: Investigation, conceptualization, methodology, data curation, and writing (original draft, review, and editing): J.L.; project administration and supervision: W.J., Y.J., and S.W.; software, experimental process, and data processing: J.L. and X.Z. All authors have read and agreed to the published version of the manuscript.

Funding: This study was supported by the Key Laboratory Research Funds of Ministry of Culture and Tourism (WHB1801).

Conflicts of Interest: The authors declare no conflict of interest.

Appendix A

The timbre materials mentioned in Section 2.1 contains 72 instruments, including 37 Chinese orchestral instruments, 11 Chinese minority instruments, and 24 Western orchestral instruments (Table A1). The names of the Chinese orchestral instruments and Chinese minority instruments are given in their original languages (Chinese), with an accompanying English translation.

Table A1. Instrument list.

Category	Type	Name of the Instrument			
Chinese Orchestral Instruments (37)	Bowed Instrument (7)	高胡 (Gaohu)	二胡 (Erhu)	中胡 (Zhonghu)	革胡 (Gehu)
		低音革胡 (Bass Gehu)	京胡 (Jinghu)	板胡 (Banhu)	
	Wind Instrument (17)	梆笛 (Bangdi)	曲笛 (Qudi)	新笛 (Xindi)	
		高音笙 (Soprano Sheng)	中音笙 (Tenor Sheng)	低音笙 (Bass Sheng)	
		高音唢呐 (Soprano Suona)	中音唢呐 (Alto Suona)	次中音唢呐 (Tenor Suona)	低音唢呐 (Bass Suona)
		高音管 (Soprano Guan)	中音管 (Alto Guan)	低音管 (Bass Guan)	倍低音管 (Doublebass Guan)
		埙 (Xun)	箫 (Xiao)	巴乌 (Bawu)	
	Plucked Instrument (10)	小阮 (Soprano Ruan)	中阮 (Alto Ruan)	大阮 (Bass Ruan)	
		柳琴 (Liuqin)	琵琶 (Pipa)	扬琴 (Yangqin)	古筝 (Guzheng)
		古琴 (Guqin)	箜篌 (Konghou)	三弦 (Sanxian)	
	Percussion Instrument (3)	编钟 (Bell chimes)	编磬 (Bianqing)	云锣 (Yunluo)	
Chinese Minority Instruments (11)	Bowed Instrument (4)	艾捷克 (Ejieke)	四胡 (Sihu)	马头琴 (Matouqin)	潮尔 (Chaoer)
	Wind Instrument (4)	朝鲜唢呐 (Chaoxian Suona)	葫芦笙 (Hulusheng)	葫芦丝 (Hulusi)	大岑 (Dacen)
	Plucked Instrument (3)	热瓦普 (Rewapu)	都塔尔 (Dutaer)	伽琴 (Gayageum)	

Table A1. *Cont.*

Category	Type	Name of the Instrument			
Western Orchestral Instruments (24)	Bowed Instrument (4)	Violin	Viola	Cello	Double bass
	Woodwind Instrument (6)	Piccolo	Flute	Oboe	Clarinet
		Bassoon	Saxophone		
	Brass Instrument (4)	Trumpet	Trombone	French horn	Tuba
	Keyboard Instrument (4)	Piano	Harpsichord	Organ	Accordion
	Plucked Instrument (1)	Harp			
	Percussion Instrument (5)	Celesta	Vibraphone	Chimes	Xylophone
		Marimba			

References

1. Chen, X. *Sound and Hearing Perception*; China Broadcasting and Television Press: Beijing, China, 2006.
2. Moore, B.C.; Glasberg, B.R.; Baer, T. A model for the prediction of thresholds, loudness, and partial loudness. *J. Audio Eng. Soc.* **1997**, *45*, 224–240.
3. Meddis, R.; O'Mard, L. A unitary model of pitch perception. *J. Acoust. Soc. Am.* **1997**, *102*, 1811–1820. [CrossRef]
4. Patel, A.D. *Music, Language, and the Brain*; Oxford University Press: Oxford, UK, 2010.
5. ANSI S1.1-1994. *American National Standard Acoustical Terminology*; Acoustical Society of America: New York, NY, USA, 1994.
6. Zwicker, E.; Fastl, H. *Psychoacoustics: Facts and Models*; Springer Science & Business Media: Berlin, Germany, 2013; Volume 22.
7. Cermak, G.W.; Cornillon, P.C. Multidimensional analyses of judgments about traffic noise. *J. Acoust. Soc. Am.* **1976**, *59*, 1412–1420. [CrossRef]
8. Kuwano, S.; Namba, S.; Fastl, H.; Schick, A. Evaluation of the impression of danger signals-comparison between Japanese and German subjects. In *Contributions to Psychological Acoustics*; Schick, A., Klatte, M., Eds.; BIS: Oldenburg, Germany, 1997; pp. 115–128.
9. Iwamiya, S.-I.; Zhan, M. A comparison between Japanese and Chinese adjectives which express auditory impressions. *J. Acoust. Soc. Jpn. (E)* **1997**, *18*, 319–323. [CrossRef]
10. Stepanek, J. Relations between perceptual space and verbal description in violin timbre. *acústica 2004 Guimarães* **2004**, 077.
11. Kim, S.; Bakker, R.; Ikeda, M. Timbre preferences of four listener groups and the influence of their cultural backgrounds. In Proceedings of the Audio Engineering Society: Audio Engineering Society Convention 140, Paris, France, 4–7 June 2016.
12. Solomon, L.N. Semantic Approach to the Perception of Complex Sounds. *J. Acoust. Soc. Am.* **1958**, *30*, 421–425. [CrossRef]
13. von Bismarck, G. Timbre of steady sounds: A factorial investigation of its verbal attributes. *Acta Acust. United Acust.* **1974**, *30*, 146–159.
14. Pratt, R.L.; Doak, P.E. A subjective rating scale for timbre. *J. Sound Vibrat.* **1976**, *45*, 317–328. [CrossRef]
15. Namba, S.; Kuwano, S.; Hatoh, T.; Kato, M. Assessment of musical performance by using the method of continuous judgment by selected description. *Music Percept.* **1991**, *8*, 251–275. [CrossRef]
16. Ethington, R.; Punch, B. SeaWave: A system for musical timbre description. *Comput. Music J.* **1994**, *18*, 30–39. [CrossRef]
17. Faure, A.; Mcadams, S.; Nosulenko, V. Verbal correlates of perceptual dimensions of timbre. In Proceedings of the 4th International Conference on Music Perception and Cognition, Montréal, QC, Canada, 14–15 August 1996.
18. Howard, D.M.; Tyrrell, A.M. Psychoacoustically informed spectrography and timbre. *Organised Sound* **1997**, *2*, 65–76. [CrossRef]

19. Shibuya, K.; Koyama, T.; Sugano, S. The relationship between KANSEI and bowing parameters in the scale playing on the violin. In *IEEE SMC'99 Conference Proceedings. 1999 IEEE International Conference on Systems, Man, and Cybernetics (Cat. No.99CH37028)*; IEEE: Tokyo, Japan, 1999; Volume 4, pp. 305–310.

20. Kuwano, S.; Namba, S.; Schick, A.; Hoege, H.; Fastl, H.; Filippou, T.; Florentine, M.; Muesch, H. The timbre and annoyance of auditory warning signals in different countries. In Proceedings of the INTERNOISE, Nice, France, 27–30 August 2000.

21. Disley, A.C.; Howard, D.M. Timbral semantics and the pipe organ. In Proceedings of the Stockholm Music Acoustic Conference 2003, Stockholm, Sweden, 6–9 August 2003; pp. 607–610.

22. Moravec, O.; Štepánek, J. Verbal description of musical sound timbre in Czech language. In Proceedings of the Stockholm Music Acoustic Conference 2003, Stockholm, Sweden, 6–9 August 2003; pp. SMAC–1–SMAC–4.

23. Collier, G.L. A comparison of novices and experts in the identification of sonar signals. *Speech Commun.* **2004**, *43*, 297–310. [CrossRef]

24. Martens, W.L.; Marui, A. Constructing individual and group timbre spaces for sharpness-matched distorted guitar timbres. In Proceedings of the Audio Engineering Society: Audio Engineering Society Convention 119, New York, NY, USA, 7–10 October 2005.

25. Disley, A.C.; Howard, D.M.; Hunt, A.D. Timbral description of musical instruments. In Proceedings of the International Conference on Music Perception and Cognition, Bologna, Italy, 22–26 August 2006; pp. 61–68.

26. Stepánek, J. Musical sound timbre: Verbal description and dimensions. In Proceedings of the 9th International Conference on Digital Audio Effects (DAFx-06), Montreal, QC, Canada, 18–20 September 2006; pp. 121–126.

27. Katz, B.; Katz, R.A. *Mastering Audio: The Art and the Science.*, 2nd ed.; Focal Press: Oxford, UK, 2007.

28. Howard, D.; Disley, A.; Hunt, A. Towards a music synthesizer controlled by timbral adjectives. In Proceedings of the 14th International Congress on Sound & Vibration, Cairns, Australia, 9–12 July 2007.

29. Barbot, B.; Lavandier, C.; Cheminée, P. Perceptual representation of aircraft sounds. *Appl. Acoust.* **2008**, *69*, 1003–1016. [CrossRef]

30. Pedersen, T.H. *The Semantic Space of Sounds*; Delta: Atlanta, GA, USA, 2008.

31. Alluri, V.; Toiviainen, P. Exploring perceptual and acoustical correlates of polyphonic timbre. *Music Percept.* **2010**, *27*, 223–242. [CrossRef]

32. Fritz, C.; Blackwell, A.F.; Cross, I.; Woodhouse, J.; Moore, B.C. Exploring violin sound quality: Investigating English timbre descriptors and correlating resynthesized acoustical modifications with perceptual properties. *J. Acoust. Soc. Am.* **2012**, *131*, 783–794. [CrossRef]

33. Altinsoy, M.E.; Jekosch, U. The semantic space of vehicle sounds: Developing a semantic differential with regard to customer perception. *J. Audio Eng. Soc.* **2012**, *60*, 13–20.

34. Elliott, T.M.; Hamilton, L.S.; Theunissen, F.E. Acoustic structure of the five perceptual dimensions of timbre in orchestral instrument tones. *J. Acoust. Soc. Am.* **2013**, *133*, 389–404. [CrossRef]

35. Zacharakis, A.; Pastiadis, K.; Reiss, J.D. An interlanguage study of musical timbre semantic dimensions and their acoustic correlates. *Music Percept.* **2014**, *31*, 339–358. [CrossRef]

36. Skovenborg, E. Development of semantic scales for music mastering. In Proceedings of the Audio Engineering Society: Audio Engineering Society Convention 141, Los Angeles, CA, USA, 29 October–1 November 2016.

37. Wallmark, Z. A corpus analysis of timbre semantics in orchestration treatises. *Psychol. Music* **2019**, *47*, 585–605. [CrossRef]

38. Chen, K.-A.; Wang, N.; Wang, J.-C. Investigation on human ear's capability for identifing non-speech objects. *Acta Phys. Sin.* **2009**, *58*, 5075–5082. [CrossRef]

39. Herrera-Boyer, P.; Peeters, G.; Dubnov, S. Automatic classification of musical instrument sounds. *J. New Music Res.* **2003**, *32*, 3–21. [CrossRef]

40. Bowman, C.; Yamauchi, T. Perceiving categorical emotion in sound: The role of timbre. *Psychomusicol. Music Mind Brain* **2016**, *26*, 15–25. [CrossRef]

41. Gupta, C.; Li, H.; Wang, Y. Perceptual evaluation of singing quality. In *2017 Asia-Pacific Signal and Information Processing Association Annual Summit and Conference (APSIPA ASC)*; IEEE: Kuala Lumpur, Malaysia, 2017; pp. 577–586.

42. Allen, N.; Hines, P.C.; Young, V.W. Performances of human listeners and an automatic aural classifier in discriminating between sonar target echoes and clutter. *J. Acoust. Soc. Am.* **2011**, *130*, 1287–1298. [CrossRef]

43. Wang, N.; Chen, K.-A. Regression model of timbre attribute for underwater noise and its application to target recognition. *Acta Phys. Sin.* **2010**, *59*, 2873–2881.

44. Blauert, J. *Communication Acoustics*; Springer: Berlin/Heidelberg, Germany, 2005; Volume 2.

45. Jensen, K. Timbre Models of Musical Sounds. Ph.D. Thesis, Department of Computer Science, University of Copenhagen, Copenhagen, Denmark, 1999.

46. Desainte-Catherine, M.; Marchand, S. Structured additive synthesis: Towards a model of sound timbre and electroacoustic music forms. In Proceedings of the International Computer Music Conference (ICMC99), Beijing, China, 22–28 October 1999; pp. 260–263.

47. Aucouturier, J.J.; Pachet, F.; Sandler, M. "The way it sounds": Timbre models for analysis and retrieval of music signals. *IEEE Trans. Multimedia* **2005**, *7*, 1028–1035. [CrossRef]

48. Burred, J.; Röbel, A.; Rodet, X. An accurate timbre model for musical instruments and its application to classification. In *Learning the Semantics of Audio Signals, Proceedings of the First International Workshop, LSAS 2006*; Cano, P., Nürnberger, A., Stober, S., Tzanetakis, G., Eds.; Technische Informationsbibliothek u. Universitätsbibliothek: Athens, Greece, 2006; pp. 22–32.

49. Wang, X.; Meng, Z. The consonance evaluation method of Chinese plucking instruments. *Acta Acust.* **2013**, *38*, 486–492.

50. Sciabica, J.-F.; Bezat, M.-C.; Roussarie, V.; Kronland-Martinet, R.; Ystad, S. Towards the timbre modeling of interior car sound. In Proceedings of the 15th International Conference on Auditory Display, Copenhagen, Denmark, 18–22 May 2009.

51. Grey, J.M. Multidimensional perceptual scaling of musical timbres. *J. Acoust. Soc. Am.* **1977**, *61*, 1270–1277. [CrossRef]

52. McAdams, S.; Winsberg, S.; Donnadieu, S.; De Soete, G.; Krimphoff, J. Perceptual scaling of synthesized musical timbres: Common dimensions, specificities, and latent subject classes. *Psychol. Res.* **1995**, *58*, 177–192. [CrossRef]

53. Martens, W.L.; Giragama, C.N. Relating multilingual semantic scales to a common timbre space, 2002. In Proceedings of the Audio Engineering Society Convention 113, Los Angeles, CA, USA, 5–8 October 2002.

54. Martens, W.L.; Giragama, C.N.; Herath, S.; Wanasinghe, D.R.; Sabbir, A.M. Relating multilingual semantic scales to a common timbre space-Part II. In Proceedings of the Audio Engineering Society Convention 115, New York, NY, USA, 10–13 October 2003.

55. Zacharakis, A.; Pastiadis, K. Revisiting the luminance-texture-mass model for musical timbre semantics: A confirmatory approach and perspectives of extension. *J. Audio Eng. Soc.* **2016**, *64*, 636–645. [CrossRef]

56. Simurra Sr, I.; Queiroz, M. Pilot experiment on verbal attributes classification of orchestral timbres. In Proceedings of the Audio Engineering Society Convention 143, New York, NY, USA, 18–21 October 2017.

57. Melara, R.D.; Marks, L.E. Interaction among auditory dimensions: Timbre, pitch, and loudness. *Percept. Psychophys.* **1990**, *48*, 169–178. [CrossRef]

58. Zhu, J.; Liu, J.; Li, Z. Research on loudness balance of Chinese national orchestra instrumental sound. In Proceedings of the 2018 national acoustical congress of physiological acoustics, psychoacoustics, music acoustics, Beijing, China, 9–12 November 2018; pp. 34–35.

59. EBU – TECH 3253. *Sound Quality Assessment Material Recordings for Subjective Tests. Users' handbook for the EBU SQAM CD*; EBU: Geneva, Switzerland, 2008.

60. Alías, F.; Socoró, J.; Sevillano, X. A review of physical and perceptual feature extraction techniques for speech, music and environmental sounds. *Appl. Sci.* **2016**, *6*, 143. [CrossRef]

61. Peeters, G. A large set of audio features for sound description (similarity and classification). *CUIDADO IST Proj. Rep.* **2004**, *54*, 1–25.

62. Peeters, G.; Giordano, B.L.; Susini, P.; Misdariis, N.; McAdams, S. The timbre toolbox: Extracting audio descriptors from musical signals. *J. Acoust. Soc. Am.* **2011**, *130*, 2902–2916. [CrossRef] [PubMed]

63. Pollard, H.F.; Jansson, E.V. A tristimulus method for the specification of musical timbre. *Acta Acust. United Acust.* **1982**, *51*, 162–171.

64. Krimphoff, J.; McAdams, S.; Winsberg, S. Caractérisation du timbre des sons complexes. II. Analyses acoustiques et quantification psychophysique. *J. Phys. IV* **1994**, *4*, C5-625–C5-628. [CrossRef]

65. Scheirer, E.; Slaney, M. Construction and evaluation of a robust multifeature speech/music discriminator. In *1997 IEEE International Conference on Acoustics, Speech, and Signal Processing*; IEEE Comput. Soc. Press: Munich, Germany, 1997; pp. 1331–1334.

66. Lartillot, O. *MIRtoolbox 1.7.2 User's Manual*; RITMO Centre for Interdisciplinary Studies in Rhythm, Time and Motion, University of Oslo: Oslo, Norway, 2019.

67. Meng, Z. *Experimental Psychological Method for Subjective Evaluation of Sound Quality*; National defence of Industry Press: Beijing, China, 2008.
68. Hodeghatta, U.R.; Nayak, U. Multiple linear regression. In *Business Analytics Using R - A Practical Approach*; Apress: Berkeley, CA, USA, 2017; pp. 207–231.
69. Yeh, C.-Y.; Huang, C.-W.; Lee, S.-J. A multiple-kernel support vector regression approach for stock market price forecasting. *Expert Syst. Appl.* **2011**, *38*, 2177–2186. [CrossRef]
70. Haykin, S.S. *Neural Networks and Learning Machines.*, 3rd ed.; Pearson education: Upper Saddle River, NJ, USA, 2009.
71. Witten, I.H.; Frank, E.; Hall, M.A.; Pal, C.J. *Data Mining: Practical Machine Learning Tools and Techniques*; Morgan Kaufmann Publishers, Inc.: San Francisco, CA, USA, 2016.
72. Borg, I.; Groenen, P.J.; Mair, P. *Applied Multidimensional Scaling*; Springer Science & Business Media: Berlin/Heidelberg, Germany, 2012.
73. Chen, K.-A. *Auditory Perception and Automatic Recognition of Environmental Sounds*; Science Press: Beijing, China, 2014.
74. Susini, P.; McAdams, S.; Winsberg, S.; Perry, I.; Vieillard, S.; Rodet, X. Characterizing the sound quality of air-conditioning noise. *Appl. Acoust.* **2004**, *65*, 763–790. [CrossRef]
75. Tucker, S. *An Ecological Approach to the Classification of Transient Underwater Acoustic Events: Perceptual Experiments and Auditory Models*; University of Sheffield: Sheffield, UK, 2003.
76. Shepard, R.N. Representation of structure in similarity data: Problems and prospects. *Psychometrika* **1974**, *39*, 373–421. [CrossRef]
77. Borg, I.; Groenen, P.J.F.; Mair, P. Variants of different MDS models. In *Applied Multidimensional Scaling. SpringerBriefs in Statistics*; Springer: Berlin/Heidelberg, Germany, 2013; pp. 37–47.
78. Borg, I.; Groenen, P. Modern multidimensional scaling: Theory and applications. *J. Educ. Meas.* **2003**, *40*, 277–280. [CrossRef]

Article

Microphone and Loudspeaker Array Signal Processing Steps towards a "Radiation Keyboard" for Authentic Samplers

Tim Ziemer [1,*,†] **and Niko Plath** [2,†]

[1]   Bremen Spatial Cognition Center, University of Bremen, 28359 Bremen, Germany
[2]   Department of Media Technology, Hamburg University of Applied Sciences, 22081 Hamburg, Germany;
      niko.plath@haw-hamburg.de
*   Correspondence: ziemer@uni-hamburg.de
†   These authors contributed equally to this work.

Received: 30 December 2019; Accepted: 23 March 2020; Published: 29 March 2020

Abstract: To date electric pianos and samplers tend to concentrate on authenticity in terms of temporal and spectral aspects of sound. However, they barely recreate the original sound radiation characteristics, which contribute to the perception of width and depth, vividness and voice separation, especially for instrumentalists, who are located near the instrument. To achieve this, a number of sound field measurement and synthesis techniques need to be applied and adequately combined. In this paper we present the theoretic foundation to combine so far isolated and fragmented sound field analysis and synthesis methods to realize a *radiation keyboard*, an electric harpsichord that approximates the sound of a real harpsichord precisely in time, frequency, and space domain. Potential applications for such a radiation keyboard are conservation of historic musical instruments, music performance, and psychoacoustic measurements for instrument and synthesizer building and for studies of music perception, cognition, and embodiment.

Keywords: microphone array; wave field synthesis; acoustic holography; sampler; synthesizer

PACS: 43.10.Ln; 43.20.+g; 43.20.Bi; 43.20.Fn; 43.20.Px; 43.20.Tb; 43.20.Ye; 43.25.Lj; 43.25.Qp; 43.28.We; 43.30.Zk; 43.38.Lc; 43.38.Md; 43.40.+s; 43.40.At; 43.40.Cw; 43.40.Dx; 43.40.Fz; 43.40.Le; 43.40.Rj; 43.40.Sk; 43.58.Jq; 43.60.+d; 43.60.Ac; 43.60.Gk; 43.60.Hj; 43.60.Lq; 43.60.Pt; 43.60.Sx; 43.60.Tj; 43.60.Uv; 43.60.Wy; 43.66.+y; 43.66.Lj; 43.75.+a; 43.75.Cd; 43.75.Gh; 43.75.Mn; 43.75.St; 43.75.Tv; 43.75.Yy; 43.75.Zz; 01.50.fd

MSC: 76Q05; 97U80

---

1. Introduction

Synthesizers tend to focus on timbral aspects of sound, which contains temporal and spectral features [1,2]. This is even true for modern synthesizers that imitate musical instruments by means of physical modeling [3,4]. Many samplers and electric pianos on the market use stereo recordings, or pseudostereo techniques [5,6] to create some perceived spaciousness in terms of *apparent source width* or *perceived source extent*, so that the sound appears more natural and vivid. However, such techniques do not capture the sound radiation characteristics of musical instruments, which may be essential for an authentic experience in music listening and musician-instrument-interaction.

Most sound field synthesis approaches synthesize virtual monopole sources or plane waves by means of loudspeaker arrays [7,8]. Methods to incorporate the sound radiation characteristics of musical instruments are based on sparse recordings of the sound radiation characteristics [5], like far field recordings from circular [9,10] or spherical [11] microphone arrays with 24 to 128 microphones. In these studies, a nearfield mono recording is extrapolated from a virtual source point. However, instead of a monopole point source, the measured radiation characteristic is included in the extrapolation function, yielding a so-called *complex point source* [9,12,13]. Complex point sources are a drastic simplification of the actual physics of musical instruments. However, complex point sources were demonstrated to create plausible physical and perceptual fields [5,14]. These sound natural in terms of source localization, perceived source extent and timbre, especially when listeners and/or sources move during the performance [5,15–20].

To date, sound field synthesis methods to reconstruct the sound radiation characteristics of musical instruments do not incorporate exact nearfield microphone array measurements of musical instruments, as described in [21–25]. This is most likely because the measurement setup and the digital signal processing for high-precision microphone array measurements are very complex on their own. The methods include optimization algorithms and solutions to inverse problems. The same is true for sound field synthesis approaches that incorporate complex source radiation patterns.

In this paper we introduce the theoretic concept of a *radiation keyboard*. We describe on a theoretical basis, and with some practical considerations, which sound field measurement and synthesis methods should be combined, and how to combine them utilizing their individual strengths. All presented results are preparatory for the realization. In contrast to conventional samplers, electric pianos, etc., a radiation keyboard recreates not only the temporal and spectral aspects of the original instrument, but also its spatial attributes. The final radiation keyboard is basically a MIDI keyboard whose keys trigger different driving signals of a loudspeaker array in real-time. When playing the radiation keyboard, the superposition of the propagated loudspeaker driving signals should create the same sound field as the original harpsichord would do. Thus, the radiation keyboard will create a more realistic sound impression than conventional, stereophonic samplers. This is especially true for musical performance, where the instrumentalists moves their heads. The radiation keyboard can serve, for example,

- as a means to produce authentic sounding replicas of historic musical instruments in the context of cultural heritage preservation [26,27],
- as an authentic and immersive alternative to physical modeling synthesizers, conventional samplers, electrical pianos (or harpsichords, respectively) [3,4,28],
- as a research tool for instrument building [29], interactive psychoacoustics [30], and embodied music interaction [31].

The remainder of the paper is organized as follows. Section 2 describes all the steps that are carried out to measure and synthesize the sound radiation characteristics of a harpsichord. In Section 2.1, we describe the setup to measure impulse responses of the harpsichord and the radiation keyboard loudspeakers. These are needed to calculate impulse responses that serve as raw loudspeaker driving signals. For three different frequency regions, f1 to f3, different methods are ideal to calculate. In Sections 2.2–2.4, we describe how to derive the loudspeaker impulse responses for frequency regions f1, f2 and f3. In Section 3, we describe how to combine the three frequency regions, and how to create loudspeaker driving signals that synthesize the original harpsichord sound field during music performance. After a summary and conclusion in Section 4, we discuss potential applications of the radiation keyboard in the outlook Section 5.

2. Method

The concept and design of the proposed radiation keyboard are illustrated in Figure 1. The sound field radiated by a harpsichord is analyzed by technical means. Then, this sound field is synthesized by the radiation keyboard. The radiation keyboard consists of a loudspeaker array whose driving signals are triggered by a MIDI keyboard. The superposition of the propagated loudspeaker driving signals creates the same sound field as a real harpsichord. To date, no sound field synthesis method is able to radiate all frequencies in the exact same way the harpsichord does. Therefore, we combine different sound field analysis and synthesis methods. This combination offers an optimal compromise: low frequencies f1 < 1500 Hz are synthesized with high precision in the complete half-space above the sound board, mid frequencies $1500 \leq$ f2 ≤ 4000 Hz are synthesized with high precision within an extended listening region, and higher frequencies f3 > 4000 Hz are synthesized with high precision at discrete listening points within the listening region.

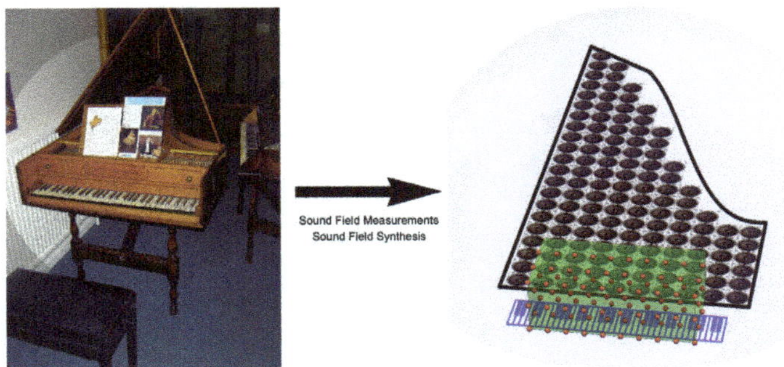

Figure 1. Design and concept of the radiation keyboard (**right**). A MIDI keyboard triggers individual signals for 128 loudspeakers, which are arranged like a harpsichord. Replacing the real harpsichord (**left**) by the radiation keyboard creates only subtle audible differences. Unfortunately, the radiation keyboard cannot synthesizes the harpsichord sound in the complete space. Low frequencies f1 are synthesized with high precision in the complete half-space above the loudspeaker array (light blue zone). Mid frequencies are synthesized with high precision in the listening region (green zone) in which the instrumentalist is located. The sound field of very high frequencies is synthesized with a high precision at discrete listening points within the listening region (red dots).

To implement a radiation keyboard, four main steps are carried out. Figure 2 shows a flow diagram of the main steps: firstly, the sound radiation characteristics of the harpsichord are measured by means of microphone arrays. Secondly, an optimal constellation of loudspeaker placement and sound field sampling is derived from impulse these measurements. As the third step the impulse responses for the loudspeaker array are calculated. These serve as raw loudspeaker driving signals. Finally, loudspeaker driving signals are calculated by a convolution of harpsichord source signals with the array impulse responses. These driving signals are triggered by a MIDI keyboard and play in real-time. The superposition of the propagated driving signals synthesizes the complex harpsichord sound field.

Figure 2. Flow diagram of the proposed method. First, harpsichord Impulse Responses (IR) are measured, then the distribution of loudspeakers and listening points is optimized, and finally, loudspeaker driving signals are calculated.

To synthesize the sound field, it is meaningful to divide the harpsichord signal into three frequency regions: frequency region f1 lies below 1.5 kHz, the Nyquist frequency of the proposed loudspeaker array. Frequency region f2 ranges from 1.5 kHz to 4 kHz, the Nyquist frequency of the microphone array. Frequency region f3 lies above these Nyquist frequencies. Different sound field measurement and synthesis methods are optimal for each region. They are treated separately in the following sections.

2.1. Setup

The setup for the impulse response measurements is illustrated in Figure 3 for a piano under construction. In the presented approach the piano is replaced by a harpsichord. An acoustic vibrator excites the instrument at the termination point between the bridge and a string for each key. Successive microphone array recordings are carried out in the near field to sample the sound field at $M = 1500$ points parallel to the sound board.

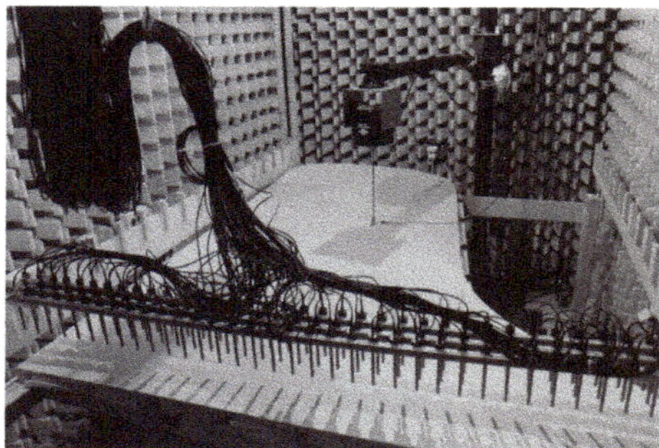

Figure 3. Measurement setup including a movable microphone array (in the front) above a soundboard, and an acoustic vibrator (in the rear) installed in an anechoic chamber. After the nearfield recordings parallel to the sound board, the microphone array samples the listening region of the instrumentalist in playing position.

In addition to the near field recordings, the microphone array samples the *listening region*. The head of the instrumentalist will be located in this region during the performance (ear channel distance to keyboard y $\approx$ 0.37 m, ear channel distance to ground z $\approx$ 1.31 m for a grown person). The location is indicated by black dots in Figure 4.

Figure 4. Depiction of the radiation keyboard. A regular grid of loudspeakers is installed on a board. The board has the shape of the harpsichord sound board. Microphones (black dots) sample the listening region in front of the keyboard.

A lightweight piezoelectric accelerometer measures the vertical polarization of the transverse string acceleration $h(\kappa_a, t)$ at the intersection point between string and bridge for each of the $A = 62$ keys. This is not illustrated in Figure 3 but indicated as brown dots in Figure 5. The acceleration measured by the sensor is proportional to the force acting on the bridge. Details on the setup can be found in [32,33]. Alternatively, $h(\kappa_a, t)$ can be recorded optically, using a high speed camera and the setup described in [34,35], or it can be synthesized from a physical plectrum-string model [36]. The string recording represents the *source signal* that excites the harpsichord.

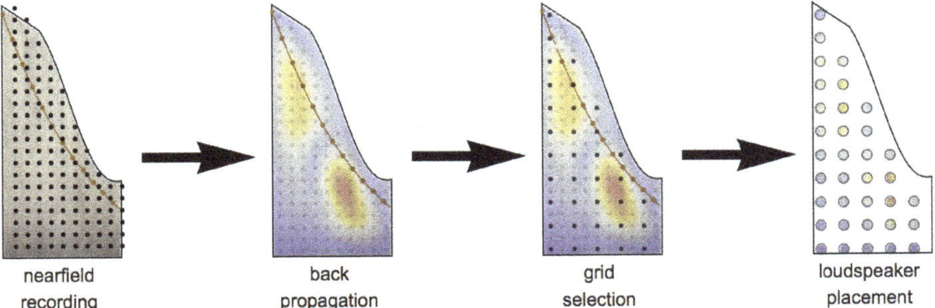

Figure 5. Procedure to derive the impulse response $R_{f1}(\Theta_l, \kappa_a, t)$ for each loudspeaker and pressed key in the radiation keyboard. The brown curve represents the bridge, the brown dots depict exemplary excitation points. The black dots represent microphones near the soundboard, the light gray dots represent equivalent sources on the soundboard. The gray dots represent a regular subset of equivalent sources, which are replaced by loudspeakers (circles) in the radiation keyboard.

Measuring string vibrations isolated from the impulse response measurements of the sound board adds a lot of flexibility to the radiation keyboard. We derive impulse responses for the loudspeaker array of the radiation keyboard. This radiates all frequencies the same way the harpsichord would do. Consequently, any arbitrary source signal can serve as an input for the radiation keyboard. In addition to the measured harpsichord string acceleration $h(\kappa_a, t)$, the radiation keyboard can load any sound sample, such as alternative harpsichord tunings, alternative instrument recordings, or arbitrary test signals.

Figure 4 illustrates the radiation keyboard. A rigid board in the shape of the harpsichord sound board serves as a loudspeaker chassis. A regular grid of loudspeakers is arranged on this chassis. The radiated sound field created by each single loudspeaker is recorded in the listening region.

2.2. Low Frequency Region f1

The procedure to calculate the loudspeaker impulse responses for frequency region f1 is illustrated in Figure 5. Firstly, impulse responses of the harpsichord are recorded in the near field. Next, the recorded sound field is propagated back to $M = 1500$ points on the harpsichord sound board. Then, an optimal subset of these points is identified. This subset determines the loudspeaker distribution of the radiation keyboard.

2.2.1. Nearfield Recording

The setup for the near field recordings is illustrated in Figure 3. A microphone array $\mathbf{X}_{near,m}$ with equidistant microphone spacing of 40 mm is installed at a distance of 5 cm parallel to the harpsichord soundboard surface. The index $m = 1, \ldots, 1500$ describes a microphone position above the harpsichord.

An acoustic vibrator excites the instrument at the intersection point of string and bridge; the *string termination point* [32] κ_a. Here, the index $a = 1, \ldots, A$ describes the pressed key. For a harpsichord with 5 octaves, $A = 62$ keys exist.

To obtain impulse responses from the recorded data the so-called exponential sine sweep (ESS) technique is utilized [37]. The method has originally been proposed for measurements of weakly non-linear systems in room acoustics (e.g., loudspeaker excitation in a concert hall) but can also be adapted to structure-borne sound [38]. For the excitation an exponential sine sweep

$$s(t) = \sin(\omega(t)), \tag{1}$$

is used, where

$$w(t) = w_1 e^{\left(\frac{t \ln\left(\frac{w_2}{w_1}\right)}{T}\right)}. \tag{2}$$

Here, $w_1 = 2\pi \text{ rad s}^{-1}$ is the starting frequency, $w_2 = 2\pi \times 24{,}000 \text{ rad s}^{-1}$ is the maximum frequency and $T = 25$ s is the signal duration. The vibrator excites the sound board with this signal. Figure 6 shows the spectrogram of an exemplary microphone recording $p(\mathbf{X}_1, t)$. Since the frequency axis has logarithmic scaling, the sweep appears as a straight line. Due to non-linearity in the shaker excitation the recording shows harmonic distortions parallel to the sinusoidal sweep.

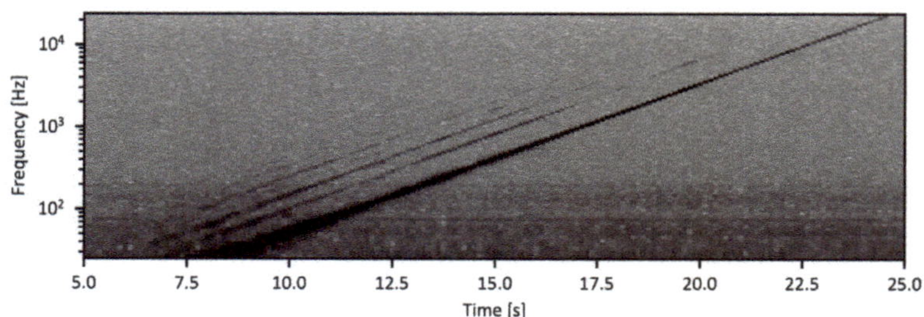

Figure 6. Spectrogram of an exemplary output of one of the array microphones. Harmonic distortions of several orders are observable.

A deconvolution process eliminates these distortions. The deconvolution is realized by a linear convolution of the measured output $p(\mathbf{X}_m, t)$ with the function

$$u(t) = s^{-1}(t)b(t). \tag{3}$$

Here, $s^{-1}(t)$ is the temporal reverse of the excitation sweep signal (2) and $b(t)$ is an amplitude modulation that compensate the energy generated per frequency, reducing the level by 6 dB/octave, starting with 0 dB at $t = 0$ s and ending with $-6\log_2(w_2/w_1)$ dB at $t = T$, expressed as

$$b(t) = 1/e^{\frac{t \ln \frac{w_2}{w_1}}{T}}. \tag{4}$$

This linear deconvolution delays $s(t)$ of an amount of time varying with frequency. The delay is proportional to the logarithm of frequency. Therefore, $s^{-1}(t)$ stretches the signal with a constant slope, and compresses the linear part to a time delay corresponding to the filter length. The harmonic distortions have the same slope as the linear part and are, therefore, also packed to very precise times. If T is large enough, the linear part of an impulse response is temporally clearly separated from the non-linear pseudo IR.

This deconvolution process yields one signal

$$q'(\mathbf{X}_m, t) = p(\mathbf{X}_m, t) * u(t). \tag{5}$$

This signal $q'(\mathbf{X}_m, t)$ is the linear impulse response $q(\mathbf{X}_m, t)$ preceded by the nonlinear distortion products, i.e., the pseudo-IRs. An example of $q'(\mathbf{X}_1, t)$ and $q(\mathbf{X}_1, t)$ is illustrated in Figure 7. The linear impulse response part can be obtained by a peak detection searching for the last peak in the time series. In the figure the final, linear impulse response $q(\mathbf{X}_1, t)$ is highlighted in red.

The driving signal and the convolution are reproducible. Repeated measurements are carried out to sample the radiated sound field. To cover each key and sample point, this yields $N \times A = 93{,}000$ recordings.

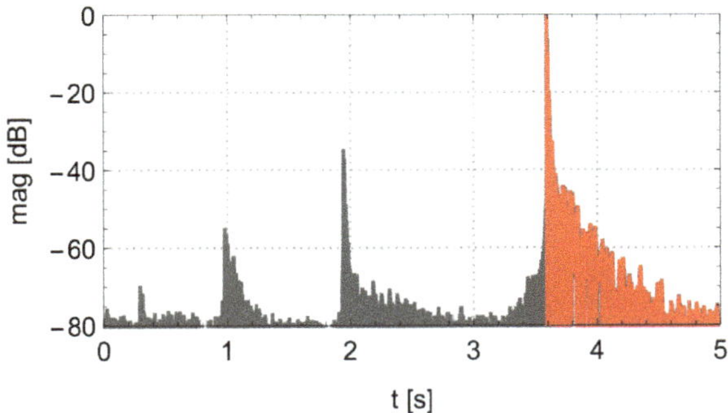

Figure 7. Obtained impulse response $q'(\mathbf{X}_m, t)$ of the signal in Figure 6 after deconvolution. The harmonic distortions are separated in time and precede the linear part $q(\mathbf{X}_m, t)$ (red), which starts at $t \approx 3.5\,\text{s}$.

2.2.2. Back Propagation

The harpsichord soundboard is a continuous radiator of sound, but can be simplified as a discrete distribution of $N = 1500$ radiating points $\mathbf{Y}_n$, referred to as *equivalent sources* [23]. These equivalent sources sample the vibrating sound board. The validity of this simplification is restricted by the Nyquist-Shannon theorem, i.e., two equivalent sources per wave length are necessary. The following steps are frequency-dependent. Therefore, we transfer functions of time into frequency domain using the discrete Fourier transform. Terms in frequency domain are indicated by capital letters and the ω in the argument. For example $Q(\omega)$ represents the frequency spectrum of $q(t)$.

The relationship between the radiating soundboard $Q(\mathbf{Y}_n, \omega)$ and the spectra of the aligned, linearized microphone recordings $Q(\mathbf{X}_m, \omega)$ is described by a linear equation system

$$Q(\mathbf{X}_m, \omega) = \sum_{n=1}^{N} G(r, \omega) Q(\mathbf{Y}_n, \omega), \tag{6}$$

where

$$G(r, \omega) = \frac{e^{i(kr)}}{r} \tag{7}$$

is the Free field Greens' function. It is a *complex transfer function* that describes the *sound propagation* of the equivalent sources as monopole radiators. Here, the term $r = ||\mathbf{X}_m - \mathbf{Y}_n||_2$ is the Euclidean distance between equivalent sources and microphones, k is the wave number and i, the imaginary unit. Equation (6) is closely related to the Rayleigh Integral which is applied in acoustical holography and sound field synthesis approaches, like wave field synthesis and ambisonics [39]. One problem with Equation (6) is that the linear equation system is ill-posed. The radiated sound $Q(\mathbf{X}_m, \omega)$ is recorded but the source sound $Q(\mathbf{Y}_n, \omega)$, which created the recorded sound pressure distribution, is sought. When solving the linear equation system, e.g., by means of Gaussian elimination or an inverse matrix of $G(r, \omega)$ [40], the resulting sound pressure levels tend to be huge due to small numerical errors, measurement and equipment noise. This can be explained by the propagation matrix being ill-conditioned when microphone positions are close to one another compared to the considered wavelength. In this case the propagation matrix condition number is high. A regularization method relaxes the matrix and yields lower amplitudes. An overview about regularization methods can be found in [21,23,40]. For musical instruments, the Minimum Energy Method (MEM) [23,41] is very powerful. The MEM is an iterative approach, gradually reshaping the radiation characteristic of $G(r, \omega)$ from monopole at $\Omega = 0$ to a ray at $\Omega = \infty$ using the formulation

$$\Psi(\alpha, \omega, \Omega) = 1 + \Omega \times (1 - \alpha), \tag{8}$$

where $\Psi(\alpha,\omega,\Omega)$ is multiplied by $G(r,\omega)$ in Equation (6) to reshaped the complex transfer function, like

$$Q(\mathbf{X}_m,\omega) = G(r,\omega) \times \Psi(\alpha,\omega,\Omega) \times Q(\mathbf{Y}_n,\omega). \tag{9}$$

In Equations (8) and (9), α describes the angle between equivalent sources $\mathbf{Y}_n$ and loudspeakers $\mathbf{X}_m$ as inner product of both position vectors

$$\alpha_{m,n} = \left| \frac{\mathbf{Y}_n \mathbf{X}_m}{|\mathbf{X}_m|\,|\mathbf{Y}_n|} \mathbf{n} \right|. \tag{10}$$

The angle α is given by the constellation of source- and receiver positions and is 1 in normal direction $\mathbf{n}$ of the considered equivalent source position and 0 in the orthogonal direction. The ideal value for Ω minimizes the reconstruction energy

$$\Omega_{\mathrm{opt}} = \min_{\Omega}\{E_\Omega\} \tag{11}$$

where

$$E_\Omega \propto \sum_n^N \left| \frac{\partial P(\mathbf{Y}_n,\omega)}{\partial \mathbf{n}} \right|^2. \tag{12}$$

The energy E is proportional to the sum of the squared pressure amplitudes on the considered structure. In a first step, the linear equation system is solved for integers from $\Omega = 0$ to $\Omega = 10$ and the reconstruction energy is plotted over Ω. Around the local minimum, the linear equation system is again solved, this time in steps of 0.1. Typically, the iteration is truncated after the first decimal place. An example of reconstruction energy over Ω is illustrated in Figure 8 together with the condition number of $G \times \Psi$ in Equation (9). Near Ω_{opt}, both the signal energy and the matrix condition number tend to be low.

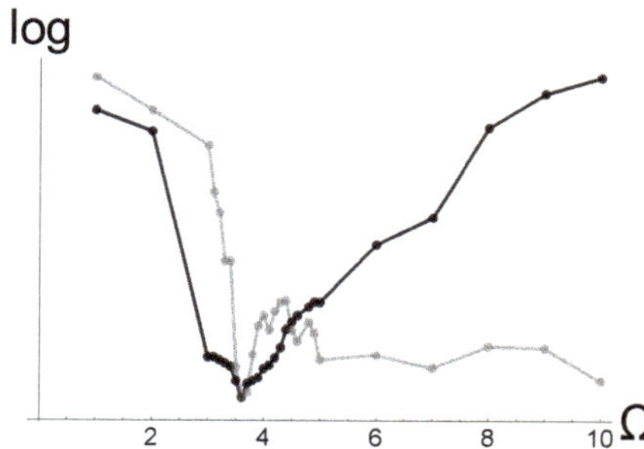

Figure 8. Exemplary plot of reconstruction energy (black) and condition number (gray) over Ω. The ordinate is logarithmic and aligned at the minimum at $\Omega_{\mathrm{opt}} = 3.6$.

Alternatively, the parameter Ω can be tuned manually to find the best reconstruction visually; the correct solution tends to create the sharpest edges at the instrument boundaries, with pressure amplitudes near 0. This is a typical result of the truncation effect: the finite extent of the source causes an acoustic short-circuit. At the boundary, even strong elongations of the sound board create hardly any pressure fluctuations, since air flows around the sound board. The effect can be observed in Figure 9.

The result of the MEM is one source term $Q(\mathbf{Y}_n,\omega)$ for each equivalent source on the harpsichord sound board. Below the Nyquist frequency, these 1500 equivalent sources approximate the sound field of a real harpsichord. This is true for the complete half-space above the sound board.

The radiation of numerous musical instruments has been measured using the described microphone array setup and the MEM, like grand piano [35,42], vihuela [23,43], guitars [43,44], drums [23,41,45], flutes [41,45], and the New Ireland kulepa-ganeg [46]. The method is so robust that the geometry of the instruments becomes visible, as depicted in Figure 9.

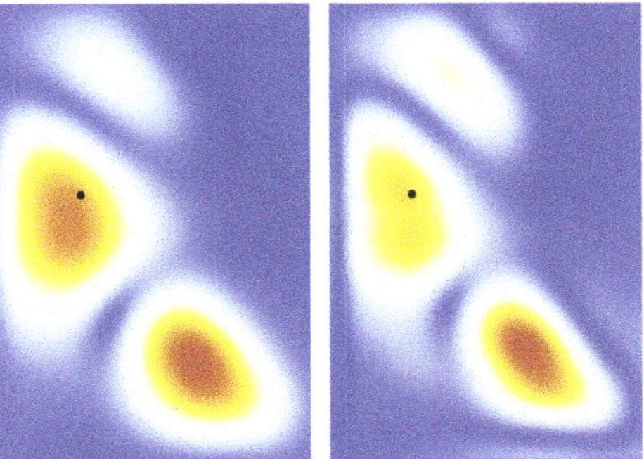

Figure 9. Sound field recorded 50 mm above a grand piano sound board (**left**) and back-propagated sound board vibration according to the MEM (**right**). The black dot marks the input location of the acoustic vibrator.

2.2.3. Optimal Loudspeaker Placement

The back-propagation method described in Section 2.2.2 yields one source term for each of the $N = 1500$ equivalent sources and $A = 62$ keys. Together, the equivalent sources sample the harpsichord sound field in the region of the sound board. Forward propagation of the source terms approximates the harpsichord sound field in the whole half-space above the sound board. This has been demonstrated, e.g., in [47]. Replacing each equivalent source by one loudspeaker is referred to as *acoustic curtain*, which is the origin of wave field synthesis [39,48]. In physical terms, this situation is a spatially truncated discrete Rayleigh integral, which is the mathematical core of wave field synthesis [7,8,39,49]. A prerequisite is that all equivalent sources are homogeneous radiators in the half-space above the soundboard. This is the case for the proposed radiation keyboard. For low frequencies, loudspeakers without a cabinet approximate dipoles fairly well. Naturally, single loudspeakers with a diameter in the order of 10 cm are inefficient radiators of low frequencies [50]. However, this situation improves when a dense array of loudspeakers is moving in phase. This is typically happening in the given scenario; when excited with low frequencies, the sound board vibrates as a whole [51], so the loudspeaker signals for the wave field synthesis will be in phase. While truncation creates artifacts in most wave field synthesis setups, referred to as *truncation error* [8,39,48], no artifacts are expected in the described setup due to natural tapering: at the boundaries of the loudspeaker array, an acoustic short-circuit will occur. However, the acoustic short-circuit also occurs in real musical instruments, as demonstrated in Figure 9. This is because compressed air in the front flows around the sound board towards the rear, instead of propagating as a wave. The acoustic short-circuit of the outermost loudspeakers acts like a natural tapering window. In wave field synthesis installations artificial tapering is applied to compensate the truncation error.

The MEM describes the sound board vibration by $N = 1500$ equivalent source terms. Replacing all equivalent sources by an individual loudspeaker is not ideal, because the spacing is too dense for broadband loudspeakers, and it is challenging to synchronize 1500 channels for real-time audio processing. Audio interfaces including D/A-converters for $L = 128$ synchronized channels in audio-cd quality are commercially available, using, for example, MADI or Dante protocol. In wave field synthesis systems, regular loudspeaker distributions have been reported to deliver the best synthesis results [52]. Covering the complete soundboard of a harpsichord with a regular grid consisting of 128 grid points, yields about one loudspeaker every 12 cm. This is a typical loudspeaker density in wave field synthesis systems and yields a Nyquist frequency of about 1.5 kHz for waves in air [7].

Three exemplary loudspeaker arrays are illustrated in Figure 10.

Every third equivalent source can be replaced by a loudspeaker with little effect on the sound field synthesis precision below 1.5 kHz. This yields $W = 11$ possible loudspeaker grid positions $\Theta'_{w,l}$.

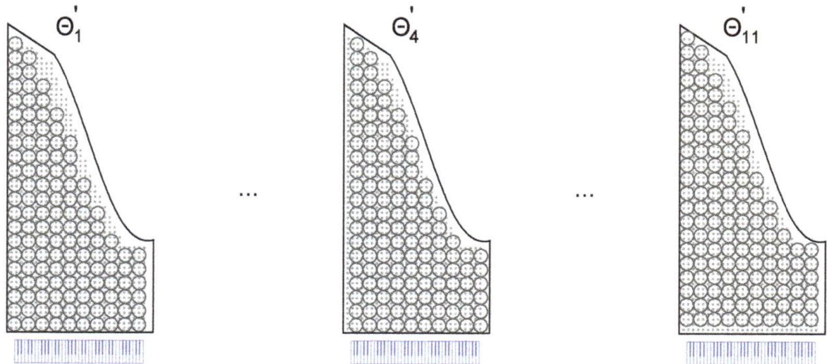

Figure 10. Three regular loudspeaker grids $\Theta'_{w,l}$. A subset of the 1500 equivalent sources is replaced by 128 loudspeakers. In the given setup $W = 11$ loudspeaker grids are possible.

At the ideal location of the loudspeaker grid, all loudspeakers lie near antinodes of all frequencies of all keys. In contrast to regions near the nodes, sound field calculations near the antinodes do not suffer from equipment noise, numerical noise, and small microphone misplacements. Instead, all loudspeakers contribute efficiently to the wave field synthesis. Therefore, the optimal grid location Θ_l has the largest signal energy

$$\Theta_l = \max_w \left\{ \sum_{a=1}^{A} \int_{t=0}^{\infty} \left(R'_{f1}(\Theta'_{w,l}, \kappa_a, t) * h(\kappa_a, t) \right)^2 \mathrm{d}t \right\}. \tag{13}$$

Equation (13) is solved for each of the $w = 1, \ldots, 11$ possible loudspeaker grid positions. The grid with the maximum signal energy is replaced by loudspeakers as indicated in Figure 5. This ideal grid is the optimal loudspeaker distribution Θ_l.

The Nyquist frequency of the loudspeaker array lies around 1.5 kHz. For reproduction of higher frequencies, other methods are necessary, as described in the following sections.

2.3. Higher Frequency Region f2

The procedure to calculate the loudspeaker impulse responses for frequency region f2 is illustrated in Figure 11. First, impulse responses of the harpsichord are recorded in the listening region. Next, impulse responses of the final loudspeaker grid Θ_l are recorded in the listening region. These are transformed such that the loudspeaker array creates the harpsichord sound field in the listening region. Then, the optimal position of listening points is determined. These listening points are a subset of the microphone locations that sample the listening region.

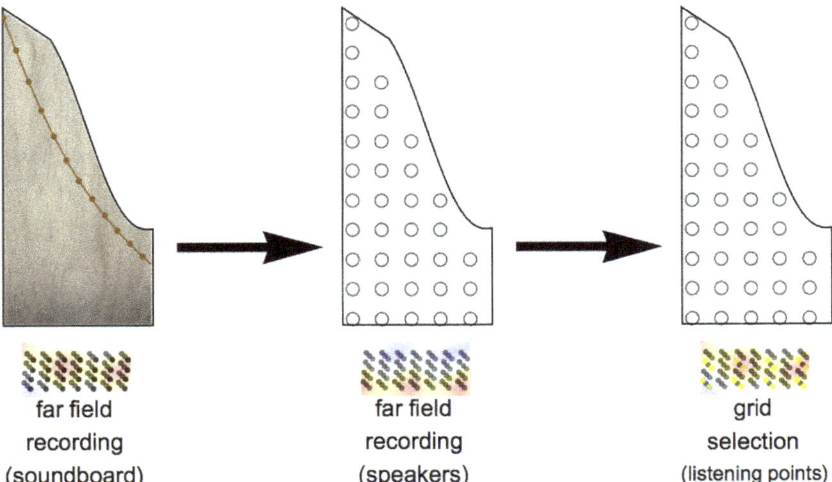

Figure 11. Procedure to derive the impulse response $R'_{f2}(\Theta_l, \kappa_a, v, t)$ for each loudspeaker Θ_l and pressed key κ_a in the radiation keyboard. The wooden plate represents the harpsichord sound board. The white plate represents the loudspeaker grid. The black dots represent microphones microphones in the listening region. At first, harpsichord and loudspeaker array create different sound fields in the listening region. Then, the loudspeaker signals are modified as to synthesize the harpsichord sound field at regular subset v of listening points that sample the listening region.

2.3.1. Far Field Recording

In addition to the near field recordings, the radiated sound is also recorded with a microphone array $X_{\mathrm{far},v,j}$ that samples the region in which the instrumentalist's head may be located during playing. We refer to this region as the *listening region* and to the discrete sample points as *listening points*. The distance between equivalent sources on the sound board and the listening points lies in the order of decimeters to meters. For frequencies above 1.5 kHz, this means that the listening region lies in the far field.

In the near field measurement, Section 2.2.1, one microphone array samples a planar region parallel to the sound board. In the far field measurement the microphone array samples a rectangular cuboid. The setup for the far field recordings is illustrated in Figures 4 and 11. As described in Sections 2.1 and 2.2.1, the sound board is excited with an exponential sweep. An array of $J = 128$ microphones samples the listening region. The microphones are arranged as a regular grid with a spacing of 4 cm. The array samples the complete sound field in the listening region for all wave lengths above 0.08 m, i.e., frequencies below 4 kHz. About $V = 11$ repeated measurements are carried out with a slightly shifted microphone array. Equations (1)–(5) describe how to excite the sound board and derive impulse responses for the $A \times J \times V = 87{,}296$ different source-receiver constellations.

These far field impulse responses provide a sample of the *desired sound field* $Q_{\mathrm{des}}(X, \kappa_a, \omega)$ in the region in which the instrumentalist is moving her head. In frequency domain, it can be described as the relationship between source signal $S(\omega)$, complex transfer function $G(r, \omega)$ and microphone array recordings $P_{\mathrm{des}}(X, \kappa_a, \omega)$

$$S(\omega) \times G(r, t) = P_{\mathrm{des}}(X_{v,j}, \kappa_a, \omega) \tag{14}$$

where the recording of the sweep is aligned to receive the impulse response

$$Q_{\mathrm{des}}(X_{v,j}, \kappa_a, \omega) = P_{\mathrm{des}}(X_{v,j}, \kappa_a, t) \times U(\omega). \tag{15}$$

The terms $S(\omega)$ and $G(r,t)$ in Equation (14) are known. So instead of microphone array measurements, $P_{\text{des}}(\mathbf{X}, \kappa_a, \omega)$ can be calculated by this forward-propagation formula. In [47] it was demonstrated that the forward-propagation equals the measurements.

2.3.2. Radiation Method

To synthesize the desired sound field $Q_{\text{des}}(\mathbf{X}, \kappa_a, \omega)$ with the given loudspeaker array Θ_l, $L = 128$ loudspeaker signals $R'_{f2}(\Theta_l, \omega)$ need to be calculated. This is done in two steps. First, the swept sine, Equation (1), is played trough each individual loudspeaker Θ_l and recorded in the listening region. Here,

$$S(\omega) \times Q(\alpha, \omega) = P_\Theta(\mathbf{X}_{v,j}, \omega)$$
$$\iff Q(\alpha, \omega) = \frac{P_\Theta(\mathbf{X}_{v,j}, \omega)}{S(\omega)} \tag{16}$$

and

$$Q_\Theta(\mathbf{X}_{v,j}, \omega) = P_\Theta(\mathbf{X}_{v,j}, \omega) \times U(\omega) \tag{17}$$

describe the relationship between the source signal, the raw microphone recordings and the final impulse response. Here, the unknown propagation term $Q(\alpha, \omega)$ is the ratio of the source signal and the recordings. In contrast to a real harpsichord source signal $H(\kappa_a, \omega)$ we know that $S(\omega) \gg 0$ for all audible frequencies. Thus, the complex transfer function $Q(\alpha, \omega)$ between each loudspeaker of the array Θ_l and each listening point $\mathbf{X}_{\text{far},v,j}$ is determined by simply recording the propagated swept sine, Equation (16), of each loudspeaker at each listening point, followed by the deconvolution, Equation (17).

This complex transfer function is neither the idealized monopole source radiation $G(r, \omega)$, nor the energy-optimized radiation function $\Psi(\alpha, \omega)G(r, \omega)$. Instead, $Q(\alpha, \omega)$ is the actual transfer function as measured physically. It includes the frequency and phase response of the loudspeakers, the amplitude decay and the phase-shift from each loudspeaker to each receiver. It can thus be considered the true transfer function. It includes the sound radiation characteristics of the loudspeakers, which tend to deviate from G and Ψ. Solving the linear equation system

$$Q_{\text{des}}(\mathbf{X}_{v,j}, \kappa_a, \omega) = R'_{f2}(\Theta_l, \mathbf{X}_{v,j}, \kappa_a, \omega) \times Q(\alpha, \omega) \tag{18}$$

for all $V = 11$ microphone array positions yields the impulse response for the loudspeakers $R'_{f2}(\Theta_l, \mathbf{X}_{v,j}, \kappa_a, \omega)$. This procedure is referred to as *radiation method* as it synthesizes a desired sound field by including the measured sound radiation characteristics of the loudspeakers. Accounting for the actual transfer function from each loudspeaker to each listening point has the advantage that the rows in the linear equation system described by Equation (18) tend to deviate stronger in reality compared to idealized monopole radiators. This has been demonstrated in [20,49]. The radiation method is a robust regularization method that has been demonstrated to relax the linear equation system [9,20,39,49]. It leads to (a) low amplitudes and (b) solutions that vary only slightly, when the source-receiver constellation or the source signal is varied slightly. The method synthesizes a desired sound field as long as (a) the sound field lies in the far field and (b) at least two listening points per wavelength exist. The method is only ideal for frequency region f2, as it does not account for nearfield effects and spatial aliasing [9,20,49].

2.3.3. Optimal Listening Points

So far, Equation (18) delivers $V = 11$ sets of $L = 128$ impulse responses for each key. The solutions only vary slightly, due to microphone misplacements, equipment and background noise, numerical errors and the spatial variations of the loudspeaker sound radiation characteristics. The solutions are valid inside the listening region. Outside the listening region synthesis errors occur, because loudspeaker signals interfere in an arbitrary manner. Outside an anechoic chamber, this will lead to unnatural reflections. Consequently, the ideal impulse response minimizes synthesis errors outside the listening region. This is achieved by selecting the impulse responses with the lowest signal energy

$$R_{f2}(\mathbf{\Theta}_l, \kappa_a, \omega) = \min_{v, \kappa_a}\{R'_{f2}(\mathbf{\Theta}_l, \kappa_a, \omega)\}, \tag{19}$$

where

$$R'_{f2}(\mathbf{\Theta}, \kappa_a, \omega) \propto \sum_{l=1}^{L} \int_{t=0}^{\infty} R'_{f2}(\mathbf{\Theta}_l, \mathbf{X}_{v,j}\kappa_a, t)^2 \mathrm{d}t. \tag{20}$$

The signal energy is the sum of all $L = 128$ loudspeaker impulse responses. In Equation (20) the signal energy of each key κ_a is calculated for each of the $v = 1, \ldots, 11$ microphone array positions. As defined in Equation (19), the final impulse response $R_{f2}(\mathbf{\Theta}, \kappa_a, \omega)$ for each specific key is the one where v creates minimal signal energy. This solution exhibits the most constructive interference inside the listening region. The solution is valid for the considered frequency region f2, i.e., between 1.5 and 4 kHz. For higher frequencies, a third method is ideal, as discussed in the following section.

2.4. Highest Frequency Region

In principle, the method to calculate the impulse response for frequency region f3 equals the method described in Section 2.3. For frequencies over 4 kHz the radiation method only synthesizes the desired sound field at the discrete listening points, but not in between [49]. For human listeners, neither the exact frequency nor the phase are represented in the auditory pathway ([39] Chapter 3). Amplitude of such high frequencies mainly contributes to the perception of brightness and the phase contributes to the impulsiveness of attack transients. Both are important aspects of timbre perception ([53] Chapter 11), ([39] Chapter 2).

Therefore, it is adequate to approximate the desired amplitudes and phases at discrete listening points by solving Equation (18). The difference between frequency regions f2 and f3 is the selection of optimal listening points. Instead of choosing the impulse response with minimum signal energy, Equation (20), the ideal impulse response for f3 is the shortest, because it exhibits the highest impulse fidelity. When convolved with a short impulse response, the frequencies of source signals stay in phase. Quite contrary, long impulse responses indicate out of phase relationships. Phase is mostly audible during transients. Consequently, the shortest impulse response is ideal, because it maintains the characteristic, steep attack transient of harpsichord notes.

Impulse responses for one loudspeaker and three different microphone array locations v is illustrated in Figure 12. The shortest impulse response can be identified visually.

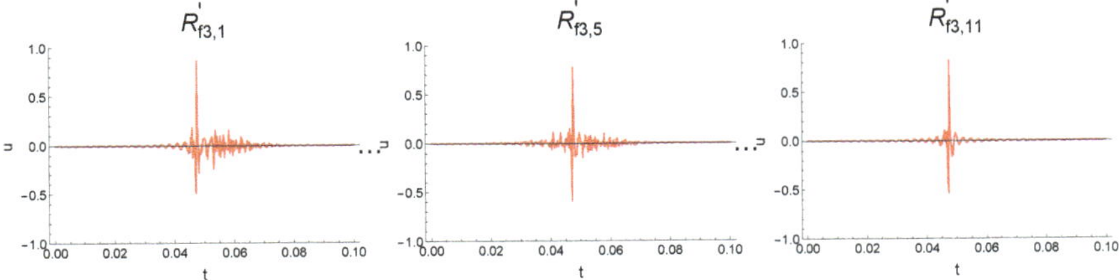

Figure 12. Three exemplary impulse responses $R'_{f3,v}$. Here, t is the time in seconds and u is the voltage to drive the loudspeaker. The shortest out of $V = 11$ impulse responses is ideal, as it exhibits the best impulse fidelity. It can be identified visually.

3. Calculation of Loudspeaker Driving Signals

Each method described above result in one truncated impulse response per loudspeaker and pressed key $R_f(\Theta, \kappa_a, t)$. Here, each impulse response is truncated in frequency. To combine them, the three truncated impulse responses per loudspeaker and pressed key are simply added, i.e.,

$$R(\Theta_l, \kappa_a, t) = R_{f1}(\Theta_l, \kappa_a, t) + R_{f1}(\Theta_l, \kappa_a, t) + R_{f1}(\Theta_l, \kappa_a, t). \tag{21}$$

The result of Equation (21) is a broadband impulse response $R(\Theta_l, \kappa_a, t)$ that covers the complete audible frequency range. An example is illustrated in Figure 13.

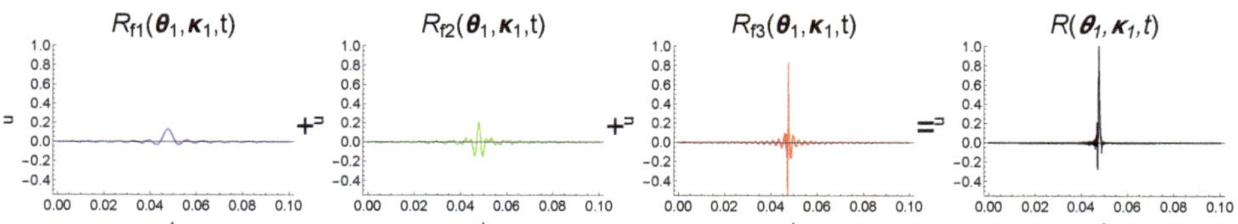

Figure 13. Impulse responses R_f truncated to frequency regions f1 (blue), f2 (green) and f3 (red). Adding up the time series yields a broadband impulse response $R(\Theta_l, \kappa_a, t)$ (black). Here, t is the time in seconds and u is a normalized voltage to drive the loudspeaker.

Summing the truncated impulse responses yields one broadband impulse response for each loudspeaker and pressed key, i.e., $B = L \times A = 7936$ broadband impulse responses $R_{\Theta_l, \kappa_a, t}$. These impulse responses describe how any frequency is radiated by the loudspeaker array to approximate the sound field that the harpsichord would create, if excited by a broadband impulse.

Naturally, a played harpsichord is not excited by an impulse. Instead, pressing a key creates a driving signal that travels through the string and transfers to the soundboard via the bridge. In Section 2.1 we provide literature that suggests three different ways to record or model this driving signal $h(\kappa_a, t)$. In order to finally use the radiation keyboard as a harpsichord sampler, the loudspeaker driving signals $d(\Theta_l, \kappa_a, t)$ are calculated by a convolution of the impulse response with the source signal, i.e.,

$$d(\Theta_l, \kappa_a, t) = h(\kappa_a, t) * R(\Theta_l, \kappa_a, t). \tag{22}$$

This yields $B = 7936$ sound files. These are imported to multiple instances of a software sampler in a digital audio workstation (DAW). Typically, one instance of a software sampler can address between 16 and 64 output channels. Consequently, between 2 and 8 sampler instances need to be initialized. Technologies like VST and Direct-X are able to handle this parallelism, and several multi-channel DAWs (like Steinberg Cubase, Ableton Live and Magix Samplitude) can handle the high number of output channels. Finally, the original keyboard of the harpsichord is replaced by a MIDI-Keyboard, whose note-on command triggers the 128 samples for the corresponding note.

As the effect of key velocity on the created level and timbre is negligibly, the harpsichord is the ideal instrument to start with; only one sample per note and loudspeaker is necessary. For more expressive instruments, such as the piano, the attack velocity affects the produced level and timbre. Here, several samples per note, or one attack-velocity controlled filter would have to be applied. This implies the need for much higher data rates and specific signal processing, which is out of scope of this paper.

4. Conclusions

In this paper the theoretic foundation of a radiation keyboard has been presented. It includes the complete chain from recording the source sound and the radiated sound of a harpsichord to synthesizing its temporal, spectral and spatial sound within an extended listening region, controlled in real-time. To achieve this, we choose the optimal method for each frequency region and inverse problem, and describe a way to combine so far isolated and fragmented sound field analysis and synthesis approaches.

For the low frequency region f1, a combination of nearfield recordings, the minimum energy method, an energy-efficient loudspeaker grid selection, and wave field synthesis is ideal. It synthesizes the desired sound field in the complete half-space above the sound board.

For the higher-frequency region f2, a combination of far field recordings, the radiation method, and energy efficient listening point selection is ideal. This combination synthesizes the desired sound field in the listening region with a high precision.

For the highest frequency region f3, far field recordings and an in-phase impulse response creation are ideal. It approximates the correct signal amplitudes in the listening region, while supporting the transient behavior of the source sound. The initial outcome of such a radiation keyboard is a sampler that mimics not only the temporal and spectral aspects of the original musical instrument, but also its spatial aspects.

5. Outlook

This paper presented the theoretic framework of our current research project. The effort to implement a radiation keyboard is very high and a number of sound field measurement and synthesis methods need to be combined, leveraging their individual strengths. We have not implemented the radiation keyboard yet; this paper rather describes the necessary means to realize it.

The implemented radiation keyboard is supposed to serve as a research tool to carry out interactive listening experiments that are more ecological than passive listening tests with artificial sounds in a laboratory environment. Note that the radiation keyboard is not restricted to harpsichord sounds. In principle, any arbitrary sound file can act as source signal and be radiated like a harpsichord. This enables us to manipulate the temporal and spectral aspects of the sound, while keeping the sound radiation constant. Loading different source sounds while keeping the sound radiation fixed, could reveal which temporal and spectral parameters affect the perception of source extent and naturalness in the direct sound of musical instruments. The radiation keyboard could answer the question, whether a saxophone sound with the radiation characteristics of a harpsichord sound larger than a real saxophone. Using the radiation keyboard, we can investigate apparent source width and immersion of direct sound both in presence and absence of room acoustics. To date, physical predictors of apparent source with originate in room acoustical investigations [15,16,39]. Findings disagree, which frequency region is of major importance for these listening impression. Different predictors and the discourse are examined in [5].

The strength of a real-time capable radiation keyboard is the interactivity: musicians can actively play the instrument instead of carrying out passive listening tests. Interactivity creates a dynamic sound and allows for a natural interaction in an authentic musical performance scenario. This is a necessity in the field of performance, gesture, and human-machine-interaction studies and a prerequisite for ecological psychoacoustics [30,31].

Author Contributions: Conceptualization, T.Z.; methodology, T.Z. and N.P.; software, T.Z. and N.P.; validation, T.Z. and N.P., formal analysis, T.Z. and N.P., investigation, T.Z. and N.P., resources, T.Z. and N.P.; data curation, T.Z. and N.P.; writing—original draft preparation, T.Z.; writing—review and editing, T.Z. and N.P.; visualization, T.Z. and N.P.; project administration, T.Z. All authors have read and agreed to the published version of the manuscript.

Funding: This research received no external funding.

Acknowledgments: We acknowledge the detailed and critical, but also amazingly constructive feedback from the two anonymous reviewers. We thank the Claussen Simon Foundation for their financial support during this project.

Conflicts of Interest: The authors declare no conflict of interest.

References

1. Beauchamp, J.W. Synthesis by spectral amplitude and "brightness" matching of analyzed musical instrument tones. *J. Audio Eng. Soc.* **1982**, *30*, 396–406.
2. Chowning, J.; Bristow, D. *FM Theory & Applications. By Musicians for Musicians*; Yamaha Music Foundation: Tokyo, Japan, 1986.
3. Bilbao, S.; Torin, A. Numerical Modeling and Sound Synthesis for Articulated String/Fretboard Interactions. *J. Audio Eng. Soc.* **2015**, *63*, 336–347. [CrossRef]
4. Pfeifle, F.; Bader, R. Real-Time Finite-Difference Method Physical Modeling of Musical Instruments Using Field-Programmable Gate Array Hardware. *J. Audio Eng. Soc.* **2016**, *63*, 1001–1016. [CrossRef]
5. Ziemer, T. Source Width in Music Production. Methods in Stereo, Ambisonics, and Wave Field Synthesis. In *Studies in Musical Acoustics and Psychoacoustics*; Schneider, A., Ed.; Springer: Cham, Switzerland, 2017; Volume 4, Current Research in Systematic Musicoogy, Chapter 10, pp. 299–340. [CrossRef]
6. Faller, C. Pseudostereophony Revisited. In Proceedings of the 118th Audio Engineering Society Convention, Barcelona, Spain, 28–31 May 2005.
7. Spors, S.; Rabenstein, R.; Ahrens, J. The Theory of Wave Field Synthesis Revisited. In Proceedings of the 124th Audio Engineering Society Convention, Amsterdam, The Netherlands, 17–20 May 2008.
8. Ziemer, T. Wave Field Synthesis. In *Springer Handbook of Systematic Musicology*; Bader, R., Ed.; Springer: Berlin/Heidelberg, Germany, 2017; Chapter 18, pp. 175–193. [CrossRef]
9. Ziemer, T.; Bader, R. Implementing the Radiation Characteristics of Musical Instruments in a Psychoacoustic Sound Field Synthesis System. *J. Audio Eng. Soc.* **2015**, *63*, 1094.
10. Albrecht, B.; de Vries, D.; Jacques, R.; Melchior, F. An Approach for Multichannel Recording and Reproduction of Sound Source Directivity. In Proceedings of the 119th Audio Engineering Society Convention, New York, NY, USA, 7–10 October 2005.
11. Zotter, F. Analysis and Synthesis of Sound-Radiation with Spherical Arrays. Ph.D. Thesis, University of Music and Performing Arts, Graz, Austria, 2009.
12. Ziemer, T.; Bader, R. Complex point source model to calculate the sound field radiated from musical instruments. *J. Acoust. Soc. Am.* **2015**, *138*, 1936. [CrossRef]
13. Ziemer, T.; Bader, R. Complex point source model to calculate the sound field radiated from musical instruments. *Proc. Mtgs. Acoust.* **2015**, *25*, 035001. [CrossRef]
14. Ziemer, T. Sound Radiation Characteristics of a Shakuhachi with different Playing Techniques. In Proceedings of the International Symposium on Musical Acoustics (ISMA-14), Le Mans, France, 7–12 July 2014; pp. 549–555.
15. Lindemann, W. Extension of a binaural cross-correlation model by contralateral inhibition. II. The law of the first wave front. *J. Acoust. Soc. Am.* **1986**, *80*, 1623–1630. [CrossRef] [PubMed]
16. Beranek, L.L. *Concert Halls and Opera Houses: Music, Acoustics, and Architecture*, 2nd ed.; Springer: New York, NY, USA, 2004.
17. Böhlke, L.; Ziemer, T. Perception of a virtual violin radiation in a wave field synthesis system. *J. Acoust. Soc. Am.* **2017**, *141*, 3875.[CrossRef]
18. Böhlke, L.; Ziemer, T. Perceptual evaluation of violin radiation characteristics in a wave field synthesis system. *Proc. Meet. Acoust.* **2017**, *30*, 035001. [CrossRef]
19. Corteel, E. Synthesis of Directional Sources Using Wave Field Synthesis, Possibilities, and Limitations. *EURASIP J. Adv. Signal Process.* **2007**, *2007*, 90509. [CrossRef]
20. Ziemer, T.; Bader, R. Psychoacoustic Sound Field Synthesis for Musical Instrument Radiation Characteristics. *J. Audio Eng. Soc.* **2017**, *65*, 482–496. [CrossRef]
21. Bai, M.R.; Ih, J.G.; Benesty, J. *Acoustic Array Systems. Theory, Implementation, and Application*; Wiley & Sons: Singapore, 2013. [CrossRef]
22. Williams, E.G. *Fourier Acoustics. Sound Radiation and Nearfield Acoustical Holography*; Academic Press: Cambridge, MA, USA, 1999.
23. Bader, R. Microphone Array. In *Springer Handbook of Acoustics*; 2nd ed.; Rossing, T.D., Ed.; Springer: Berlin/Heidelberg, Germany, 2014; pp. 1179–1207. [CrossRef]
24. Kim, Y.H. Acoustic Holography. In *Springer Handbook of Acoustics*; Rossing, T.D., Ed.; Springer: New York, NY, USA, 2007; Chapter 26, pp. 1077–1099. [CrossRef]

25. Hayek, S.I. Nearfield Acoustical Holography. In *Handbook of Signal Processing in Acoustics*; Havelock, D., Kuwano, S., Vorländer, M., Eds.; Springer: New York, NY, USA, 2008; Chapter 59, pp. 1129–1139. [CrossRef]

26. Plath, N. 3D Imaging of Musical Instruments: Methods and Applications. In *Computational Phonogram Archiving*; Bader, R., Ed.; Springer International Publishing: Cham, Switzerland, 2019; Volume 5, Current Research in Systematic Musicology, pp. 321–334. [CrossRef]

27. Pillow, B. Touching the Untouchable: Facilitating Interpretation through Musical Instrument Virtualization. In *Playing and Operating: Functionality in Museum Objects and Instruments*; Cité de la Musique—Philharmonie de Paris: Paris, France, 2020; p. 8.

28. Russ, M. *Sound Synthesis and Sampling*, 3rd ed.; Focal Press: Burlington, MA, USA, 2009.

29. Bader, R.; Richter, J.; Münster, M.; Pfeifle, F. Digital guitar workshop—A physical modeling software for instrument builders. In *Proceedings of the Third Vienna Talk on Music Acoustics*; Mayer, A., Chatziioannou, V., Goebl, W., Eds.; University of Music and Performing Arts Vienna: Vienna, Austria, 2015; p. 266.

30. Neuhoff, J.G. (Ed.) *Ecological Psychoacoustics*; Elsevier: Oxford, UK, 2004.

31. Lesaffre, M.; Maes, P.J.; Leman, M. (Eds.) *The Routledge Companion to Embodied Music Interaction*; Routledge: New York, NY, USA, 2017. [CrossRef]

32. Plath, N.; Pfeifle, F.; Koehn, C.; Bader, R. Driving Point Mobilities of a Concert Grand Piano Soundboard in Different Stages of Production. In Proceedings of the 3rd Annual COST FP1302 WoodMusICK Conference, Barcelona, Spain, 7–9 September 2016; pp. 117–123.

33. Plath, N.; Preller, K. Early Development Process of the Steinway & Sons Grand Piano Duplex Scale. In *Wooden Musical Instruments Different Forms of Knowledge*; Pérez, M.A., Marconi, E., Eds.; Cité de la Musique—Philharmonie de Paris: Paris, France, 2018; pp. 343–363.

34. Plath, N. High-speed camera displacement measurement (HCDM) technique of string vibrations. In Proceedings of the Stockholm Music Acoustics Conference (SMAC), Stockholm, Sweden, 30 July–3 August 2013; pp. 188–192.

35. Bader, R.; Pfeifle, F.; Plath, N. Microphone array measurements, high-speed camera recordings, and geometrical finite-differences physical modeling of the grand piano. *J. Acoust. Soc. Am.* **2014**, *136*, 2132. [CrossRef]

36. Perng, C.Y.J.; Smith, J.; Rossing, T. Harpsichord Sound Synthesis Using a Physical Plectrum Model Interfaced With the Digital Waveguide. In Proceedings of the 14th International Conference on Digital Audio Effects (DAFx-11), Paris, France, 19–23 September 2011; pp. 329–335.

37. Farina, A. Simultaneous measurement of impulse response and distortion with a swept-sine technique. In Proceedings of the 108th Audio Engineering Society Convention, Paris, France, 19–22 February 2000.

38. Boutillon, X.; Ege, K. Vibroacoustics of the piano soundboard: Reduced models, mobility synthesis, and acoustical radiation regime. *J. Sound Vib.* **2013**, *332*, 4261–4279. [CrossRef]

39. Ziemer, T. *Psychoacoustic Music Sound Field Synthesis*; Springer: Cham, Switzerland, 2020; Volume 7, Current Research in Systematic Musicology. [CrossRef]

40. Bai, M.R.; Chung, C.; Wu, P.C.; Chiang, Y.H.; Yang, C.M. Solution Strategies for Linear Inverse Problems in Spatial Audio Signal Processing. *Appl. Sci.* **2017**, *7*, 582. [CrossRef]

41. Bader, R. Reconstruction of radiating sound fields using minimum energy method. *J. Acoust. Soc. Am.* **2010**, *127*, 300–308. [CrossRef] [PubMed]

42. Plath, N.; Pfeifle, F.; Koehn, C.; Bader, R. Microphone array measurements of the grand piano. In *Seminar des Fachausschusses Musikalische Akustik: "Musikalische Akustik Zwischen Empirie und Theorie"*; Mores, R., Ed.; Deutsche Gesellsch. f. Akustik: Berlin, Germany, 2015; pp. 8–9.

43. Bader, R. Radiation characteristics of multiple and single sound hole vihuelas and a classical guitar. *J. Acoust. Soc. Am.* **2012**, *131*, 819–828. [CrossRef] [PubMed]

44. Bader, R. Characterizing Classical Guitars Using Top Plate Radiation Patterns Measured by a Microphone Array. *Acta Acust. United Acust.* **2011**, *97*, 830–839. [CrossRef]

45. Bader, R.; Münster, M.; Richter, J.; Timm, H. Microphone Array Measurements of Drums and Flutes. In *Musical Acoustics, Neurocognition and Psychology of Music*; Bader, R., Ed.; Peter Lang: Frankfurt am Main, Germany, 2009; Volume 25, Hamburger Jahrbuch für Musikwissenschaft, Chapter 1, pp. 15–55.

46. Bader, R. Computational Music Archiving as Physical Culture Theory. In *Computational Phonogram Archiving*; Bader, R., Ed.; Springer International Publishing: Cham, Switzerland, 2019; Volume 5, Current Research in Systematic Musicology, pp. 3–34. [CrossRef]

47. Richter, J.; Münster, M.; Bader, R. Calculating guitar sound radiation by forward-propagating measured forced-oscillation patterns. *Proc. Meet. Acoust.* **2013**, *19*, 035002. [CrossRef]
48. Ahrens, J. *Analytic Methods of Sound Field Synthesis*; Springer: Berlin/Heidelberg, Germany, 2012. [CrossRef]
49. Ziemer, T. Implementation of the Radiation Characteristics of Musical Instruments in Wave Field Synthesis Application. Ph.D. Thesis, University of Hamburg, Hamburg, Germany, 2016. [CrossRef]
50. Mitchell, J. Loudspeakers. In *Handbook for Sound Engineers*, 4th ed.; Ballou, G., Ed.; Elsevier: Burlington, MA, USA, 2008.
51. Le Moyne, S.; Le Conte, S.; Ollivier, F.; Frelat, J.; Battault, J.C.; Vaiedelich, S. Restoration of a 17th-century harpsichord to playable condition: A numerical and experimental study. *J. Acoust. Soc. Am.* **2012**, *131*, 888–896. [CrossRef] [PubMed]
52. Corteel, E. On the Use of Irregularly Spaced Loudspeaker Arrays for Wave Field Synthesis,Potential Impact on Spatial Aliasing Frequency. In Proceedings of the 9th International Conference on Digital Audio Effects (DAFx-06), Montreal, QC, Canada, 18–20 September 2006; pp. 209–214.
53. Bader, R. *Nonlinearities and Synchronization in Musical Acoustics and Music Psychology*; Springer: Berlin/Heidelberg, Germany, 2013; Volume 2, Current Research in Systematic Musicology. [CrossRef]

MDPI

St. Alban-Anlage 66

4052 Basel

Switzerland

Tel. +41 61 683 77 34

Fax +41 61 302 89 18

www.mdpi.com

Applied Sciences Editorial Office

E-mail: applsci@mdpi.com

www.mdpi.com/journal/applsci